HACKING IMMORTALITY

HACKING IMMORTALITY

New Realities in the Quest to Live Forever

From the Minds of
Sputnik Futures

Tiller Press
New York London Toronto Sydney New Delhi

An Imprint of Simon & Schuster, Inc.
1230 Avenue of the Americas
New York, NY 10020

First Tiller Press trade paperback edition April 2020

TILLER PRESS and colophon are trademarks of Simon & Schuster, Inc.

For information about special discounts for bulk purchases, please contact Simon & Schuster Special Sales at 1-866-506-1949 or business@simonandschuster.com.

The Simon & Schuster Speakers Bureau can bring authors to your live event. For more information or to book an event, contact the Simon & Schuster Speakers Bureau at 1-866-248-3049 or visit our website at www.simonspeakers.com.

Interior design by Jennifer Chung

Manufactured in the United States of America

10 9 8 7 6 5 4 3 2 1

Library of Congress Cataloging-in-Publication Data has been applied for.

ISBN 978-1-9821-3983-4
ISBN 978-1-9821-3984-1 (ebook)

To the past, present, and future visionaries who
fearlessly explore the frontier of longevity.

"**Hello,** *I am Alice, and I am always in a state of 'wander.'"*

Alice in Futureland is a book series that asks you to wander into possible, probable, plausible, provocative futures.

Consider this book a guide.

Inside, you will discover extraordinary ideas: a cross-pollination of art, science, and culture. Alice's aim is to give the future a platform for expression, so *everyone* can make sense of it–and help create it.

When speculating about the future, it's easy to get lost in the volume of information. That's where this book comes in. Alice is designed to break the static flow with a dynamic reading experience, where experimentation and exploration meet.

The ultimate purpose of the Alice series is to foster curiosity.

To enliven our present.

To be accessible to everyone.

To allow for exploration.

And to incite optimism.

So, wheeeeeeeeeee, down the rabbit hole we go!

CONTENTS

INTRODUCTION

Is It Simply a Design Flaw That We Age and Die?

Most books or articles on longevity would start at the beginning, offering a historical perspective on humanity's quest for the fountain of youth. But we actually think today is the true beginning.

It is 2020, and we are about to press "play" on several therapies and solutions that will slow and potentially reverse aging. When we look back one hundred or two hundred years from now, we believe history will show that this was the inflection point in cracking the code of radical human longevity.

We are entering the Fourth Biological Revolution, which will dramatically alter the "fourth stage" of life: eighty and beyond. Scientists used to call this the "red zone," where entropy kicks in and we spiral toward death; but now they're looking at the probability of an extended "health span," the part of life when a person is alive and healthy. This revolution is focused on achieving (near) immortality as biology becomes our ultimate technology in defeating death.

Scientists are playing with a new toolbox of biochemical processes, transcribing and manipulating our unique genetic codes to turn back the years. And we can now determine our "GrimAge," the closest known estimate of our true age (and likely expiration date).

There are dozens of "immortalists" who are researching or investing in the race to end death—or, at the very least, end aging. The cofounders of Google created their own longevity lab called Calico. PayPal cofounder Peter Thiel and Amazon founder Jeff Bezos have a vested interest in Unity Biotechnology, one of a few biotech labs looking at the cellular cause of aging. One of the world's leading geneticists, George Church, cofounded the start-up Rejuvenate Bio, which aims to eventually help humans live to 130 years of age—in the body of a twenty-two-year-old.

Biohackers are engineering longevity cocktails based on key compounds like nicotinamide mononucleotide (NMN), a vitamin B_3 metabolite that promotes healthy function of longevity genes, neurobiotics that stimulate brain function, and supplements. Technologists and transhumanists are tapping artificial intelligence and biological implants to prepare for digital immortality.

In the past thirty years, there have been many great breakthroughs in science, technology, and medicine. We have moved away from the mechanistic approach to genetics to understand that the genome is much more fluid in nature, capable of being greatly influenced by things like nutrition, environment, and emotional state. This has sparked a greater sense of personal responsibility for our health. And our cultural interest in longevity is working alongside science as we learn our biomarkers and attend to our emotional wellness, embracing a more holistic model of aging.

Let's pause a moment on the word "immortality." Some skeptics believe nothing is immortal. Science tells us that some plant and animal species reach "biological immortality," holding on to their youthful characteristics until the unfortunate time of decay. Digital eternalists believe we can upload our minds for future generations to download and chat with us. Others believe our consciousness is immortal and are bent on coding the entire neural network of the human brain into some form of immortal cosmic "dust" to rocket into outer space, where it will roam our galaxy for eternity. Whatever your take, we at Alice in Futureland like to think of immortality not as a fixed goal but rather an organic process we are striving to perfect.

Hacking Immortality ventures through the looking glass of our pursuit to live forever—or at least, as Ray Kurzweil says, "to live long enough to live forever." Think of this as a visit with genius—the mavericks, optimists, pragmatists and visionaries—who are stretching the biological boundaries of mortality in a quest to enable greater health, longer life, and prosperity.

There is no better time for failure than now. Some of these ideas are experiments, some are even moon shots, but all, in the end, will move us closer to a longer, healthier, more energetic life. Yes, there will be ethical, social, geopolitical, financial, and environmental challenges to such a radical change. But while we humans create problems, we also solve them.

Arthur C. Clarke's Three Laws of Forecasting:[1]

- When a distinguished but elderly scientist states that something is possible, he is almost certainly right. When he states that something is impossible, he is very probably wrong.
- The only way of discovering the limits of the possible is to venture a little way past them into the impossible.
- Any sufficiently advanced technology is indistinguishable from magic.

This book, like science, doesn't have the one answer or solution, but it maps the possibilities available today, and in the very near future, so that we can begin to shape our own destiny. Perhaps in one hundred or two hundred years from now, culture will look back and reflect on this moment of discovery, trials, successes, and failures, and realize this was the new dawn of humanity. Hopeful? Yes. Because the future always comes; we can't stop it. But few people participate in the dialogue. My advice to you: Participate. The optimist in us believes we are in for an exciting ride.

There's a New Disease Called "Aging": Do You Have It?

Flipping the Evolutionary Script from Life Span to Health Span

In 2013, *Time* ran a provocative cover with the headline "Can Google Solve Death?" Fast-forward a decade, and the longevity movement is reflected everywhere: in the movies we make, the buildings we design, and of course, our own biology.

In this chapter we will introduce some of the pioneers, dreamers, and innovators who are reshaping the narrative of aging. But in order to leap to the future, we must start at the beginning. Why? Because you were once a single cell. And that one cell contained all the biological intelligence you'd ever need.

Aging today is already treatable with a combination of several drugs that treat age-related diseases. A 2019 article in *AARP* magazine titled "Can a Single Pill Keep You Healthy to 100? (We May Soon Find Out)" explored RTB101, an investigational drug developed by the biopharmaceutical company resTORbio, which has been in clinical trials for the past fifteen years. The drug halts aging at a cellular level. Clinically available drugs like rapamycin and metformin, and supplements like Basis (produced by Elysium Health), are showing us that we can stop or radically slow the clock. We've reached the tipping point in aging science.

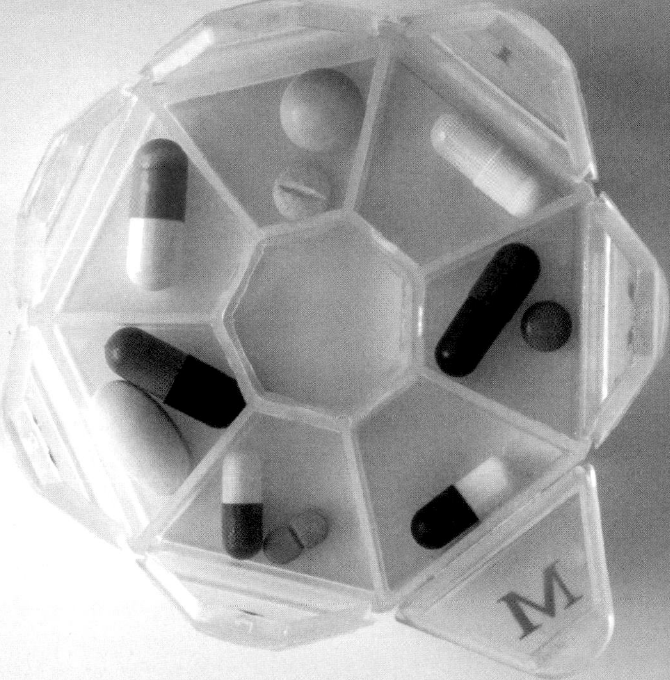

In 2013 a *Time* cover asked "Can Google Solve Death?" when Google first incubated the California Life Company, aka Calico. Now under the wing of Google's Alphabet and headed by the former chief scientist of Genentech, the company is focused on unraveling the genetics of aging and developing therapeutic solutions.

—calicolabs.com

Can you imagine living in a world where aging is optional?

The World Health Organization, in their international disease codebook, declared aging a treatable condition. So now doctors and countries can report back to the World Health Organization how many people in their country are suffering from this condition known as "old age."

—David A. Sinclair, genetics professor at Harvard Medical School, author of *Lifespan: Why We Age—and Why We Don't Have To*

It turns out that both "disease" and "aging" have arbitrary definitions. Aging is a complex process characterized by an accumulation of damage, loss of function, and increased vulnerability. Thus, one way to "treat" it is by delaying the classic age-related diseases, including cancer, heart disease, Alzheimer's, and neurodegeneration.

The World Health Organization defines "health" as "a state of complete physical, mental and social well-being, not merely the absence of infirmity or disease." In the research paper "It's Time to Classify Biological Aging as a Disease," Sven Bulterijs of Ghent University and his colleagues also note that disease is a complex phenomenon and argue that an accurate definition must consider both biological and social explanations. A new generation of researchers in gerontology seems to agree; some suggest aging needs its own code in the *International Statistical Classification of Diseases and Related Health Problems* (*ICD*). By officially recognizing aging as a disease, we may finally get the funding for new

The World Is Rapidly Aging

IN 2015 THE NUMBER OF people aged sixty-five and above stood at an estimated 617 million—8.5 percent of the global population—and this number is set to grow to 1 billion by 2030 for the first time ever, representing 12 percent of the global population. Ten billion people will be living on earth by 2050, and two billion of them will be over the age of sixty; that's 20 percent of the global population.

—"How Can We Make Healthcare Fit for the Future?" World Economic Forum, January 2016

ACCORDING TO THE CENTERS FOR Disease Control and Prevention 2013 report *The State of Aging and Health in America*, two factors—longer life spans and aging baby boomers—will combine to double the population of Americans aged sixty-five years or older during the next twenty years to about 72 million. By 2030, older adults will account for roughly 20 percent of the US population. During the past century a major shift occurred in the leading causes of death for all age groups, including older adults, from infectious diseases and acute illnesses to chronic diseases and degenerative illnesses.

—Elysium Health[1]

TODAY, 96 PERCENT OF INFANTS born in developed nations will live to age fifty and older, more than 84 percent will survive to age sixty-five or older, and 75 percent of all deaths will occur between the ages of sixty-five and ninety-five years old. Clinicians, scientists, and public health professionals should proudly declare victory in their efforts to extend the human life span to its very limits, according to University of Illinois at Chicago epidemiologist S. Jay Olshansky. He argues that the focus should shift to compressing the "red zone"—the time at the end of life characterized by frailty and disease, and extending the "health span"—the length of time when a person is alive and healthy.

drugs to fight the causes of our decline.

David A. Sinclair is one of the most prolific researchers in the area. He believes that aging is the root cause for most of the diseases we suffer from. Along with fifteen colleagues from Harvard, MIT, and other institutions around the US and Europe, he launched the nonprofit Academy for Health and Lifespan Research "to promote future work, ease collaborations between scientists, and ensure that governments and corporations are making decisions based on the latest facts instead of rumor, speculation, or hype."[2] One of the more powerful governing organizations they lobby is the Food and Drug Administration, which currently has not specifically acknowledged aging as a disease.

17
16
15
02
01

The emphasis is no longer on extending life. Most [of us] are trying to understand how to live a healthier life.

—Ilaria Bellantuono, codirector of the Healthy Lifespan Institute, University of Sheffield, UK, quoted in Anthony King, "Can We Live Forever?" *Chemistry World*, January 2019

Some of us today are already treating aging as a "pre-disease" with the promise of rejuvenating creams, supplements, and therapies. But the science coming online will give you more than hope in a bottle.

Mikhail V. Blagosklonny, researcher in cell stress biology at Roswell Park Comprehensive Cancer Center in Buffalo, New York, argues that "aging is not programmed," and several leading scientists believe that we can make our aging *programmable*. They're encouraging us to think not in terms of life span but *health span*: the period of your life when you are healthy, and free of disease.

But we first have to deal with the biology of evolution. According to biologist Michael R. Rose, aging is evolution giving up on us. At around the age of fifty-five, the force of natural selection declines—and so does aging. Who knew, right?

55theses.org is a blog dedicated to Michael R. Rose's "55 Theses on the Power and Efficacy of Natural Selection for Sustaining Health." At the heart of the challenge of the fifty-five theses is this idea that most of our health is not dependent on the health institutions but on evolved biology. If we fit our lives closely to our evolutionary design, then we will age well.

—55theses.org

The force of natural selection starts off when you're young, and it's very powerful. As you get older, through adulthood, it steadily falls, until in human populations, like in human males, by the age of fifty or fifty-five, it's zero. And here's the cool thing: it stays at zero forever after. It doesn't get any worse than that.

–Dr. Michael R. Rose, professor of ecology and evolutionary biology, University of California, Irvine, Sputnik Futures interview, 2011

Dr. Rose and his colleague Laurence D. Mueller simulated the consequences of aging for the evolution of populations like human populations, and their work showed that the foundation of aging isn't a physiological process; it is natural selection giving up on you. It's the actual dropping of the force of natural selection that generates the aging. The falling of the force of natural selection comes to an end, and it doesn't continue dropping. Rose and Mueller's theory suggests that when natural selection stops dropping, after a time lag that depends on some genetic details, aging stops. Rose has been working on this theory since the late seventies, proposing that if you change the timing of the force of natural selection–that is, change when reproduction starts–then you can make aging evolve. At later ages, adaptation responds more slowly, when natural selection is weaker. Rose's recent research proposes that as we evolve in age, so should our diets. He suggests that today we have evolved to adopt an agricultural diet, based on our current food systems and farming, which works in our earlier life as we are growing. But around age fifty-five or later, we should revert back to a paleo-ancestral-type diet, which through simulations with generations of flies has shown to have a greater health benefit in slowing aging.[3]

The Lesson of Evolution Is That There's Constantly Room for Improvement

A huge fraction of what we're going to die from in industrialized nations are diseases that don't kill twenty-year-olds. But probably 90 percent of us will die of such age-related diseases. And if you get some gene therapies that get multiple ones at once, then you're probably on the right track for something that's dealing with the fundamentals of aging rather than just alleviating symptoms . . . Or possibly, in some cases, reversing some of the disease state as well.

–George Church, in conversation with Ramez Naam in "How to Turn Science Fiction into Science Fact," Neo.Life, June 2018

Want to Be Sixty-Five in the Body of a Forty-Year-Old? You May Be Able to Make It Happen

So far, the medical community's focus has been on the DNA that guides cell replacement. A pivotal year was 2008, when two strong studies broke new ground. A study in *Nature Genetics* found that only a few genes determine whether a plant lives a single life cycle, has an annual life cycle, or endures as a perennial–and that these genes can be manipulated. A study in the *Proceedings of the National Academy of Sciences of the United States of America* detected a strong correlation between unusual human longevity and a genotype called *FOXO3a*, known as the "longevity gene." These studies, and more, offer hope that there will soon be drugs that can modulate aging, with a few US-based start-ups moving into clinical trials.

Some of these new drugs draw on one of the oldest tools in the book: caloric restriction, or CR. In recent years, scientists have discovered that we have nutrient sensors in our cells that determine food availability, and these sensors regulate large gene programs

BEFORE WE EXPLORE THE NEW drugs that draw on the pathways for antiaging through caloric restriction, such as mTOR and sirtuins, there are some ways that a strict diet alone can influence the genes that affect the aging process. Valter Longo, director of the Longevity Institute at the University of Southern California's Leonard Davis School of Gerontology and one of *Time*'s 50 Most Influential People in Health Care of 2018, has developed ProLon, a five-day "fasting-mimicking diet." ProLon's vegetarian and natural ingredients are designed to stave off hunger while tricking the body into fasting mode. It has been trialed in more than five hundred patients with metabolic syndrome; used by one hundred thousand patients in Australia, the UK, the US, and Italy; and prescribed by more than fifteen thousand doctors.

HOW IT WORKS: ProLon's five-day fasting diet helps to remove amino acids and sugars that cause master regulator genes to switch on stress resistance. "Everything is turned on to minimize damage. Meanwhile, lots of things get broken down. This includes mitochondria, proteins, and organelles in the cell," says Longo.[4] In this breakdown, the cell recognizes it can't sustain itself, so it starts to downsize and rebuild. Caloric restriction triggers this biological process of autophagy, which is the destruction of damaged or redundant cellular components. In other words, our cellular junk gets cleared away. According to Longo, they have shown that cycles of starving and feeding can regenerate the pancreas.

that, in turn, regulate how fast we age. Studies have shown CR to be effective in extending life span and deflecting age-related chronic diseases in a variety of species, including mice, fish, flies, worms, and yeast. Rodent studies conducted over the past twenty years have demonstrated up to a 40 percent increase in maximum life span through lifelong CR. The mechanisms through which this occurs are still not validated in humans, for many reasons—like, who wants to starve for twenty years?—but CR has reduced metabolic rate and oxidative stress, improved insulin sensitivity, and altered neuroendocrine and sympathetic nervous system function in animals. Most important, these studies demonstrate how it changes genes.

(Full disclosure here: Alice in Futureland are animal lovers and don't support extraneous animal testing of any kind. But our mammalian, insect, bacteria and plant counterparts are invaluable in helping scientists understand how we can co-opt the effects of therapies like caloric restriction. More on that in chapter 03.)

You, a Fruit Fly, and a Gene Called "Indy"

Not all the credit for understanding the effects of caloric restriction goes to skinny mice. There's a gene found in fruit flies that, when altered, can extend—in fact, double—the flies' life span.

Researchers dubbed the gene Indy (for "I'm not dead yet," a well-known comic line from *Monty Python and the Holy Grail*). The protein encoded by this gene transports and recycles metabolic by-products. But occasionally defects in the gene lead to production of a protein that renders metabolism less efficient, so that the fruit fly's body thinks and functions as if it were fasting, even though its eating habits are unchanged. This defect appears to create a metabolic state that mimics caloric restriction.[5]

The study of caloric restriction has produced several breakthrough compounds and clinical drugs. The first are little white pills, called Basis, sold by Elysium Health as supplements for your cells. Much like a multivitamin, Basis offers you a daily dose of "cellular health & optimization." It's basically a cocktail of powerful compounds that activate the mechanisms to slow down aging in individual cells, working on cellular proteins called sirtuins.

SOME OF THE EARLIEST RESEARCH into sirtuins and the ability to slow aging was led by Dr. Leonard Guarente and his lab at MIT, who found that these proteins are most active in cells during times of stress or limited food. Guarente's lab discovered ways to activate the sirtuins without having to practice daily fasting. And with this knowledge, a supplement was born. In 2014, Guarente cofounded Elysium Health to sell Basis direct to the public, bypassing FDA approval, which could take years.

THE BASIS PILL CONTAINS PTEROSTILBENE, a polyphenol similar to red wine's resveratrol, and nicotinamide riboside, a precursor for the coenzyme nicotinamide dinucleotides (NAD+) that helps increase available nicotinamide adenine dinucleotides (NADs) in the blood. NADs play an important role in signaling and regulatory pathways. Together, these compounds are shown to kick-start cellular sirtuin production. The company's science is backed by an impressive advisory board, including eight Nobel laureates. The Mayo Clinic and Harvard each have a financial interest in Elysium Health.

ANOTHER RESEARCHER CREDITED WITH DISCOVERING the role of sirtuins and the molecule NAD+ in disease and aging is David Sinclair, who also appears on the patent application that Elysium Health licenses for Basis. Sinclair recently shared on LinkedIn what he takes daily to treat aging: 750 milligrams of NMN every morning, along with a gram of resveratrol and 500 milligrams of metformin. So far, so good!

How a Drug for Diabetes May Become the First Antiaging Drug

My resting heart rate is fifty-seven—I checked it again this morning. That, I'm told, is the heart rate of an athlete, and I'm no athlete. My lungs operate at levels doctors expect to see among adults in their twenties—quite a shock since I've inherited the genetic lung defect that contributed to my mother's death. My LDL cholesterol and blood pressure are both considered very healthy for a *young* adult. When I exercise with weights, I recover quickly—just like I did when I was in my twenties. When I run I get bored long before I get tired.

—David A. Sinclair, "This Is Not an Advice Article," published on LinkedIn Pulse, June 25, 2018

Let's break down his recipe:

Nicotinamide mononucleotide (NMNs) are a biosynthetic precursors of nicotinamide adenine dinucleotide (NADs), known to promote cellular NAD+ production and counteract age-associated pathologies.

Resveratrol, a polyphenol that acts like antioxidants to protect the body against damage, has been popular for years. It's found in things like grapes and red wine, although you would need to consume massive amounts of a good cabernet sauvignon to get any benefits.

The new kid on the block, metformin, is a widely taken oral therapy for type 2 diabetes that is on the World Health Organization's list of essential medicines.

Metformin is turning out to be a winner in the race against aging. A drug currently dispensed to treat diabetes, it's also being tested for its life-extending (and disease-preventing) capabilities.

A UK Prospective Diabetes Study (UKPDS) compared metformin, which has been on the market in the UK since 1958, with other anti-diabetes drugs and found a decreased risk of cardiovascular disease, one of the leading age-related diseases. While it has been widely used in the UK and Canada (since 1972), the US Food and Drug Administration didn't approve metformin until 1994. Studies since then have shown that diabetics on the medication have a 17 percent lower mortality rate than people without diabetes, even though they are technically sicker. There is a randomized controlled trial underway called Targeting Aging with Metformin, or TAME, which is the first FDA-sponsored study of whether a specific drug can slow aging. The trial is being led by Nir Barzilai, director of the Institute for Aging Research at the Albert Einstein College of Medicine, and if the study succeeds in showing that metformin does regulate aging and its diseases (beyond diabetes), it would pave the way for development of next-generation drugs that directly target the biology of aging.[6]

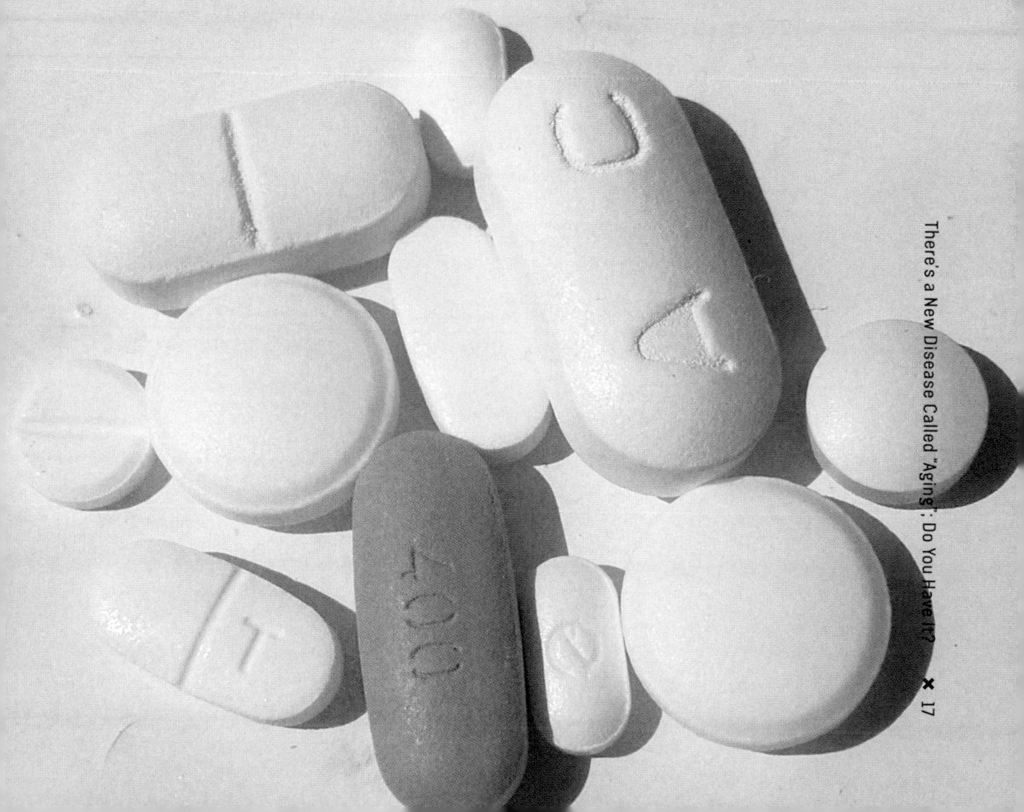

Survival of the Fittest in a Pill

Another disease-treating drug currently in clinical trials has much more exotic origins. Called rapamycin, it's an antifungal compound made by bacteria found living in the soil on the remote Easter Island. (Rapa Nui is the native name for the island.) The bacteria secreted rapamycin into the soil in order to stop the growth of competitive fungi and to absorb as many nutrients as possible themselves. Rapamycin utilizes these survival instincts with immunosuppressive effects and is used to prevent organ rejection in transplant patients. Its key mechanism is the mTOR—short for "mammalian target of rapamycin"—a protein-signaling pathway that is a master controller of cell growth metabolism.

Matt R. Kaeberlein, a pathology professor at the University of Washington School of Medicine in Seattle, believes metformin and rapamycin are two drugs with promising implications. In mice trials, both extended life span. Some of Kaeberlein's work shows that rapamycin mimics the effects of caloric restriction. The mTOR pathway is nutrient-sensitive: if you impair the pathway, the body will prepare for a situation with less nutrients and go into "repair, maintenance, rebuild" mode, similar to the impact of caloric restriction.

In the US the Dog Aging Project is treating middle-aged dogs with rapamycin to see if it also improves their health and longevity. Recent scientific findings, including those published in the scientific journals *Cell*, *Nature*, and *Science*, suggest that aging and aging-related conditions such as immunosenescence (the decline of the body's immune system) contribute to random cellular wear and tear, and to specific intracellular signaling pathways, including the mTOR pathway.

Think of the mTOR pathway as a micro-network of protein signaling, with two multi-proteins, TORC1 and TORC2, in charge of keeping the conversation going. TORC2 is the optimist and TORC1 the pessimist; preclinical studies have shown that when you inhibit TORC1 from talking too much, multiple aging-related conditions improve, including immunosenescence.

resTORbio, Inc., a clinical-stage biopharmaceutical company developing innovative medicines that target the biology of aging to prevent or treat aging-related diseases, is in a global phase 3 program to evaluate the safety and efficacy of RTB101, an oral TORC1 inhibitor. resTORbio believes it has the potential to reduce the percentage of people age sixty-five and older with clinically symptomatic respiratory illness.

Preserving your immune system could be the first step in deterring your decline. As we age, particularly beyond sixty-five, our immune systems grow tired, less able to detect and fight illnesses such as respiratory tract infections.

MANY STUDIES ARE PROVING THAT preventing and potentially reversing the effects of aging on the immune system may potentially slow down the aging process itself.

NOVARTIS INSTITUTES FOR BIOMEDICAL RESEARCH in Massachusetts put 264 volunteers aged sixty-five and over on a six-week course of two drugs designed to block mTOR. The double-blind, randomized study has shown promising results, safely reducing infections in around 40 percent of the elderly volunteers.

AROUND 80 PERCENT OF THE US health budget is spent on treating people over sixty-five years old. As more clinical trials on rapamycin and metformin prove that it can help humans live longer, and the drugs are remarketed as antiaging with FDA approval, the possibilities are endless. If people begin taking these drugs in early middle age, we may see an overall savings in health care—and the promise of significantly extending our health span.

The (37.2 Trillion) Cell Situation

A team of researchers from the Department of Experimental, Diagnostic and Specialty Medicine at the University of Bologna recently came up with the average number of cells in the human body—37.2 trillion—and that doesn't count the tens of trillions of microbes that live on and in you. You are made up of 50 billion fat cells, 35 billion skin cells, and roughly 2 billion heart muscle cells, and each of these has to keep running to keep you active, healthy, and living. Everything from bone repair to skin regrowth to the spread of infections to aging happens at the cellular level.

At the beginning of the twentieth century, scientists believed that our cells were basically immortal, continuing to divide until we died. The truth—that our cells themselves age—didn't come to light until 1961, when microbiologist Leonard Hayflick discovered that normal human cells divided only a limited number of times—forty to sixty—before stopping. This became known as the Hayflick Limit. Hayflick also discovered that normal cells have a memory and are aware of the doubling level they have reached. That division life cycle of cells is important, not only for growth, repair, and defense functions, but for understanding and fighting disease and, in turn, delaying aging.

Since our cells have a shelf life, we need to keep their engines going. That's the mitochondria's job. Mitochondria are the organelles that convert oxygen and food into ATP, or energy. Mitochondria lose steam as we age, and when they don't function properly, it can cause damage to DNA and proteins.

If We Could Push the Hayflick Limit Further, Could We Slow Down Aging and Even Cheat Death?

By focusing on the mitochondria, researchers have made some breakthrough discoveries about how to reenergize age-damaged cells that will become the basis for anti-degeneration drugs. In a study published in the journal *Nature* in 2018, scientists were able to reverse wrinkled skin and hair loss in mice by restoring the function of their mitochondria. Research led by Johan Auwerx, MD and PhD, at the École Polytechnique Fédérale's Laboratory for Integrated and Systems Physiology, in Lausanne, Switzerland, is looking at how diet, exercise, and drug intervention might positively affect mitochondrial function. In 2017, Auwerx's lab made headlines with a groundbreaking study that used the nicotinamide adenine dinucleotide-boosting compound nicotinamide riboside to improve mitochondrial function in a mouse model of Alzheimer's disease. The NAD+ reduced amyloid plaques—clumps that destroy connections between nerve cells—and improved brain function, suggesting that the disease of aging might be metabolic in nature.

AUWERX'S LAB ALSO STUDIES THE effect mitochondrial function has on proteins and how age-related diseases attack both. Proteins make up about 50 percent of your cells and carry out all the work and biochemical processes that happen inside of them. It's proteins that provide the structure and movement of the cells to keep them functioning. As you age, proteins are modified by exposure to things like high glucose. So, if the proteins become fatigued, the mitochondria become dysfunctional and the whole biochemical system suffers, which leads to aging.

High-Intensity Interval Training (HIIT) for Your Cells

The good news is that we can address both mitochondrial function and proteostasis (the process that regulates proteins within the cell) with lifestyle changes, primarily exercise.

In the 1960s, fitness pioneer Jack LaLanne preached the power of exercise to care for your 640 muscles and to fight disease and aging. Fast-forward to 2017, and a team of researchers at the Mayo Clinic published a study proving that exercise—and in particular high-intensity interval training such as biking—gives your cells a youthful boost, causing them to make more proteins for their energy-producing mitochondria and their protein-building ribosomes, effectively stopping cellular aging.

The study consisted of thirty-six men and thirty-six women from two age groups (eighteen- to thirty-year-olds and sixty-five- to eighty-year-olds) and enrolled them in three different exercise programs: high-intensity interval biking, strength training with weights, and one that combined strength training and interval training. They then took biopsies from the volunteers' thigh muscles and analyzed the molecular makeup of their muscle cells, compared to samples from sedentary volunteers. The results, published in *Cell Metabolism* in 2017, were outstanding: While strength training was effective at building muscle mass, high-intensity interval training yielded the biggest benefits at the cellular level. The younger volunteers in the interval training group saw a 49 percent increase in mitochondrial capacity, and the older volunteers saw an even more dramatic 69 percent increase. The high-intensity biking regimen also rejuvenated the volunteers' ribosomes, which are responsible for producing our cells' protein building blocks. The researchers also found a great increase in mitochondrial protein synthesis, enhanced mitochondrial function, and increased muscle.

"Based on everything we know, there's no substitute for these exercise programs when it comes to delaying the aging process," said study senior author K. Sreekumaran Nair, a medical doctor and diabetes researcher at the Mayo Clinic in Rochester, Minnesota. "These things we are seeing cannot be done by any medicine."[7]

The 5:2 and Cellular Cleaning

It's common wisdom that eating a healthy diet can keep you fit and youthful. But what it's really important for is mitochondrial health: helping clean the junk out of our cells. Too much processed or calorie-dense food leads to mitochondrial dysfunction and obesity, which is associated with a variety of other age-related health issues. Eating healthy foods and staying active also support autophagy, the process of clearing out the weakest and sickest cells in your body, including the damaged proteins, organelles, and the mitochondria that have become sluggish. Without autophagy, your body would gradually fill up with junk DNA and dead tissue.

Autophagy is critical to preventing cancer, slowing down aging, and building muscle in response to exercise. Animal studies suggest that fasting can supercharge this cellular cleansing, which has prompted biohackers and nutritionists alike to tout the benefits of fasting diets. The 5:2 diet, also known as the Fast Diet, is the most popular intermittent fasting diet. It gets its name because you eat normally five days of the week, while the other two you restrict calories to five hundred to six hundred per day.

While there are many types of fasting diets, they can all be categorized into three groups based on the timing of the fast: (1) time-restricted feeding, which is the practice of limiting calorie intake to a certain time period, usually between eight and twelve hours per day; (2) intermittent calorie restriction, which calls for reducing daily caloric intake to eight hundred to one thousand calories for two consecutive days per week; and (3) periodic fasting, which requires limiting caloric intake for between three and five days so that cells deplete glycogen stores and begin ketogenesis.

But we should note that autophagy doesn't happen all the time, and it's not designed to. We need it particularly in response to certain forms of cancer and tumor growth. While autophagy concentrates on the recycling of damaged subcellular components, like organelles, proteins, and cytoplasm, there are entire cells that sometimes become irreparably damaged. This happens simply because the cells are too old or have replicated too many times, a phenomenon known as cellular senescence. And scientists are working on how we might train our bodies to delete these old cells, commonly called "zombie" cells.

Autophagy and Physical Fitness

Autophagy plays a role in exercise, both during and after. During exercise, our cells require what's called microautophagy, the process that maintains glucose homeostasis and amino acid reserves within muscle cells. Without autophagy, our cells would quickly run out of energy. Indeed, mice genetically modified for reduced autophagy show a decrease in endurance during bouts of exercise.

Ways of Inducing Autophagy

To some degree, autophagy is always happening in the body. But as we've seen, most people would benefit from more autophagy. There are several ways to induce more autophagy, including fasting, ketogenic dieting, exercise, and drugs.

How to Use This Information

Autophagy can be a double-edged sword, but on balance it's a good thing. In short, you want more autophagy throughout your body most of the time. Unless you have cancer, your focus should be on increasing autophagy on an intermittent basis, giving your body time to recover in between bouts.

A few practical guidelines:

- Fast for twelve to sixteen hours at least once a week, and maybe as often as every day.
- Fast for around thirty-six hours at least a few times a year, and no more than once a week.
- Exercise several times a week for at least an hour, and no more than two hours.
- Exercise each individual muscle group at least twice a week for at least twenty minutes, and no more than an hour.
- Exercise each muscle group to the point of exhaustion at least once a week. In other words, deplete that muscle's glycogen stores.
- Eat at a caloric deficit for most of the year, except when specifically trying to build muscle.
- Spend a few months out of the year in a state of ketosis by following a ketogenic diet.
- Maybe take resveratrol for part of the year, cycling on and off it every so often. However, the full effects and optimal dosing protocol are unknown.

YOU DON'T NEED TO DO all of the above, but you should be doing at least two of them—ideally, one of the diet habits and one of the exercise habits.

—John Fawkes,
Los Angeles-based
personal trainer, online
fitness and nutrition coach,
and health and fitness writer,
editor of Superhuman by
Science, articles on
evidence-based fitness,
biohacking, nootropics,
and health
Medium[8]

We Need to Clean Out the Junk in Our Cells

What we propose is that we can keep people healthy late in life by repairing all of the damage that the body does to itself throughout life in the course of its normal operation. And some of that damage is . . . essentially, the accumulation of waste products both inside cells and also in the spaces between cells. But some of this is not quite like that. For example, sometimes we simply have too many of a particular bad type of cell that is misbehaving, or, in other cases, we don't have enough cells of a particular good type. Cells die, and they are not necessarily replaced automatically by the division of other cells. Furthermore, there is damage to the structure, the lattice of proteins that holds the body together, which is called the extracellular matrix. So, as you can see, there are many different types of damage, and we have to fix them all. Now, our work at SENS Research Foundation is exactly that. We are focusing on many of these things because we know that they all need to be developed and applied to people at the same time in due course. And, in a few cases, the rest of the world is already understanding this and focusing well on them. So, for example, of course, there's a huge amount of stem cell research going on worldwide. But the reason we created the Foundation was because there are plenty of other areas which were very neglected. Luckily, we've made lots of progress in that, so now, everyone's getting much more excited about it.

–Aubrey de Grey, antiaging pioneer, chief science officer, and cofounder of SENS [Strategies for Engineered Negligible Senescence] Research Foundation, from an interview on *SophieCo*, RT News, December 2018

JAMES KIRKLAND, DIRECTOR OF the Kogod Center, commented on the Mayo Clinic's website that "it's important to emphasize that, while some measurable improvement was noted in all the participants, this is simply the start of human studies. We don't know what lies ahead."

BUT IT'S A GOOD START, and others are validating it. The peer-reviewed open-access scientific journal *Nature Communications* has reported positive testing on human cells in the lab, confirming that the senolytic drugs targeted senescent cells while leaving "good" cells alone.[9]

THERE'S STILL MUCH TO LEARN about senescent cells and the long-term impact of senolytic drugs to rid the body of them. The journal *Nature Medicine* reported that treatment with a combination of two senolytic drugs, dasatinib plus quercetin, could prevent cell damage, delay physical dysfunction and extend life span in naturally aging mice.[10]

MEDICAL XPRESS REPORTS SOME MOUSE studies have shown a more direct tie between zombie cells and aging. When drugs targeting those cells were given to aged mice, the animals showed better walking speed, grip strength, and endurance on a treadmill. When the treatment was applied to very old mice, the equivalent of people ages seventy-five to ninety, it extended life span by an average of 36 percent.[11]

We Need to Defeat Our Zombie Cells

Researchers are testing several new strategies to clear senescent cells, the old cells that stop dividing but hang around. These zombie cells can neither kill themselves nor resume normal function; they just accumulate in our tissue, interfering with the functions of healthy cells, causing symptoms such as inflammation, osteoporosis, diabetes, muscle weakness, and occasionally cancer.

The new class of medicine currently in trials is called senolytics. The team at Mayo Clinic's Robert and Arlene Kogod Center on Aging has had some success destroying senescent cells in mice using senolytic agents and have recently begun human trials. Collaborating with researchers from Wake Forest School of Medicine and the University of Texas Health Science Center at San Antonio, they conducted a short safety trial on the removal of senescent cells from a small group of patients with pulmonary fibrosis to determine if it was in fact feasible to move ahead with actual large-scale human trials, and the results were promising. The study reports,

"While lung function, clinical test outcomes, frailty levels and overall health among the patients did not change, all fourteen participants showed clinically meaningful improvement in physical function in nine doses over three weeks. That ability was measured in four tests: gait speed, walking speed in six minutes, a chair rise test and a score related to a bank of physical function tests."[12]

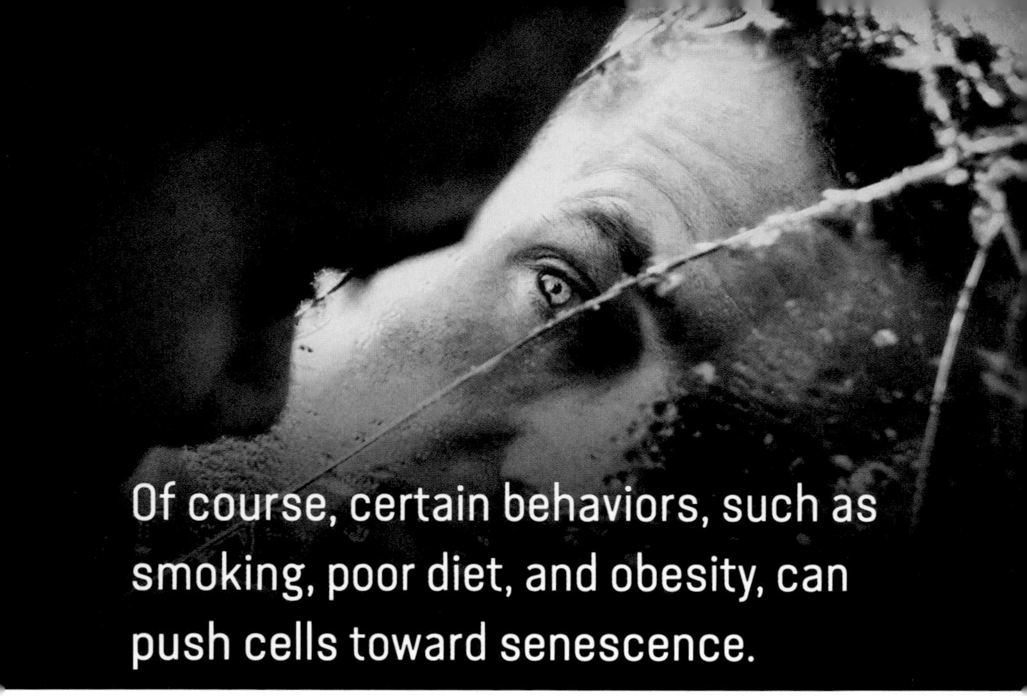

Of course, certain behaviors, such as smoking, poor diet, and obesity, can push cells toward senescence.

Judith Campisi, a scientist at the Buck Institute for Research on Aging who pioneered senescence research, refers to zombie cells as "a smoking gun." When chronically present, they promote chronic inflammation, "which is a major contributor to every major age-related disease," says Campisi. Senescent cells tend to accumulate with age, and, even worse, they drag their neighbors down with them.[13]

One thing we have yet to learn is how to disrupt zombie cells' strong communication skills. When researchers transplanted the cells into young mice, it basically made them act older: the maximum walking speed of the young mice slowed down, and their muscle strength and endurance decreased. Tests showed that the implanted cells had converted other cells to zombie status.

Campisi continues her mission to tame these persistent cells using senolytic agents. In 2018 she cofounded Unity Biotechnology, based in San Francisco, to build on work from Campisi's lab and others. In his article "Finally, the Drug That Keeps You Young" for *MIT Technology Review*, Stephen S. Hall asked Campisi: "How specifically does senescence contribute to aging?"

> The correct way to think about senescence is that it's an evolutionary balancing act. It was selected for the good purpose of preventing cancer—if [cells] don't divide, [they] can't form a tumor. It also optimizes tissue repair. But the downside is if these cells persist, which happens during aging, they can now become deleterious. Evolution doesn't care what happens to you after you've had your babies, so after around age fifty, there are no mechanisms that can effectively eliminate these cells in old age. They tend to accumulate. So the idea became popular to think about eliminating them, and seeing if we can restore tissues to a more youthful state.
>
> —Judith Campisi, PhD, professor at the Buck Institute for Research on Aging, senior scientist at Lawrence Berkeley National Laboratory, and cofounder of Unity Biotechnology

The science community is most excited about senolytics because the studies so far have shown that they can improve our health span. If the numerous trials underway with senolytic agents can show it to be effective and safe against age-related diseases, then there is hope that they could be used to fight biological aging itself. "The dream is to develop these drugs to take every few years, to clean out your senescent cells and to restore tissue function," says Campisi.

Aging happens at the cellular level. By the time symptoms become visible, it's too late. Senolytics are on track to fix that.

Unity Biotechnology is developing medicines to target vulnerabilities unique to aging cells, and to eliminate them from the human body while leaving healthy cells unaffected. The Mayo Clinic and Jeff Bezos are among the VC backers.

–UnityBiotechnology.com

The Once-Immortal Stem Cells

Now that you've met the lazy senescent cells, time to meet the superwoman stem cells. You may know about them already: there isn't a miracle face cream on the market today without them. Or you may have heard about the controversial use of placenta (embryonic) stem cells to treat many life-threatening diseases and advance regenerative medicine research. When we understand why adult stem cells age, we can understand their impact on diseases and life span.

Basically, stem cells have no bias: they can literally transform into another form of cell. The Mayo Clinic staff describes their wizardry best: "Under the right conditions in the body or a laboratory, stem cells divide to form more cells called daughter cells. These daughter cells either become new stem cells (self-renewal) or become specialized cells (differentiation) with a more specific function, such as blood cells, brain cells, heart muscle cells or bone cells. No other cell in the body has the natural ability to generate new cell types."[14]

There are three types of stem cells: adult stem cells, embryonic (or pluripotent) stem cells, and induced pluripotent stem cells (iPSCs). Scientists originally believed that adult stem cells were pickier than the others, staying within their tissue family; so, for instance, an adult liver stem cell can become an adult heart stem cell, but not a muscle stem cell. But that view has changed recently, with emerging evidence suggesting that adult stem cells may be able to create various types of cells. Adult stem cell transplants are used to replace cells damaged by chemotherapy or disease, or to stimulate the immune system to fight some types of cancer and blood-related diseases such as leukemia and lymphoma.

Embryonic (or pluripotent) stem cells are the "blank slates"–the real masters of shape-shifting, as they have the ability to literally grow into any cell type in the body without exception. Known as the "master builders," they can be the single source of thousands of different cell types.

But, alas, they are not immortal, as we once thought. In a review for the Company of Biologists, David Sinclair–yes, the same guy who gave his daily ritual for longevity–and Michael B. Schultz, a postdoctoral fellow who works with Sinclair in the Harvard Program in Therapeutic Science, explain the recent discovery that stem cells are susceptible to damage accumulation and even "zombification," just like other cells.

Are stem cells our best hope against aging?

According to the organizers of RAADfest (the annual conference on longevity produced by the Coalition for Radical Life Extension), "of all the antiaging modalities that have surfaced in recent years, stem cell therapies have received some of the most intense scrutiny"—and the most favorable attention. Is it warranted? The leading stem cell researchers believe the answer to that question is an unequivocal yes. "Stem cell clinics continue to share new evidence showing stem cell therapies have extensive antiaging benefits, from decreased inflammation and tissue regeneration to increases in physical strength and even mental clarity . . . As more human trials are conducted, stem cell therapies are showing promise across a broad range of applications for autism, heart failure, osteoarthritis, and rheumatoid arthritis."[15]

Adding New Parts: Reprogramming Stem Cells

The third type of stem cell is the one that will figure most prominently in future therapies. Known as "induced pluripotent stem cells," these are regular adult cells that have been altered, via genetic reprogramming, to give them the properties of embryonic stem cells. The same reprogramming can be used to, for instance, take regular connective tissue cells and reprogram them to become functional heart cells. The Mayo Clinic reports that studies on animals with heart failure that were injected with new heart cells experienced improved heart function and survival time. In some cases, when people get stem cell transplants for bone marrow reconstitution, if the donor is younger than the recipient, the overall biology of that recipient improves. Inspired by this phenomena, a former surgeon named Bob Hariri founded Celularity, which aims to harvest stem cells from human placentas and inject them into older people. Hariri believes this will help the frail organ systems of older individuals to improve because they will, in effect, be composed of younger cells.

We're developing the technologies to not just delay these diseases of aging but actually reverse aspects of them. Imagine you have a treatment for heart disease, but as a side effect you'd also be protected against Alzheimer's, cancer, and frailty. You'd live a longer and healthier life.

–David A. Sinclair, genetics professor and director of the Paul F. Glenn Center for the Biology of Aging at Harvard Medical School[16]

What We Can Do Right Now to Increase Vitality and Reduce Aging Biomarkers

STEP 1 is mTOR inhibition using rapamycin, which animal studies suggest can extend life span, improve cardiac function, and decrease cancer incidence, as well as suppress toxic senescent cell secretions. Recent human trials in the elderly have shown enhanced immune function.

STEP 2 is NAD+ restoration. By age eighty, levels of NAD+ will have dropped to only 4 percent. NAD+ is essential for DNA repair in each cell of the body, and any cell rejuvenation therapy, such as resveratrol or the infusion of young plasma or stem cells, is going to work better in people with optimal levels of NAD+. Once levels have been restored, they can be maintained by supplementation of nicotinamide riboside.

STEP 3 is elimination of senescent "zombie" cells and toxic debris from the cells to extend healthy life span. As cells reach the end of their life cycle or become damaged, most self-destruct by apoptosis; however, when they fail to self-eliminate, the zombie cells linger, causing inflammation and damage to surrounding healthy tissues. Senolytic therapy, including the dasatinib and quercetin regimen, is recommended in a September 2019 FightAging.org article on senolytic treatment.[17]

Another method suggested for purging senescent cells is fasting, although significant age reversal benefits may not begin until after at least day three of fasting, which may not be practical for many.

STEP 4 is the infusion of young plasma or stem cells, again stressing the importance of having completed the other steps first to eliminate inflammation and prepare the body to receive optimal benefit from these options.

—People Unlimited Inc.[18]

Now That We Know the Disease of Aging, Time to Check Your Expiration Date!

Culture has long been obsessed with death, and mythology has personified it as the "grim reaper." Who invented the grim reaper is not clear, but many believe it was the Greeks, based on the god Chronos, known as "Father Time."

What's Your GrimAge?

Predicting Death Through Your Biological Clocks

For whom does the bell toll? Just ask Steve Horvath, professor of human genetics and biostatistics at UCLA, and inventor of the "Horvath clock." He can let you know your "GrimAge," or when your clock will stop.

And that clock is in the DNA molecule. What triggers disruptions or improvements in our clocks are our genes and how well they interact with our environments and lifestyles. In fact, scientists have identified a "longevity gene" among centenarians, and one day a specific protein in the gene that reverses cardiovascular aging may dramatically increase anyone's chance of becoming a centenarian.

But why stop at one hundred? Researchers at the University of California, Berkeley, have captured the most detailed images to date of telomerase, the enzyme that lengthens the ends of chromosomes and plays a critical role in aging. Every time a cell divides, the chromosomes are copied, but they aren't copied all the way to the end. Consequently, telomeres gradually get shorter. In fact, we lose around fifty to two hundred base pairs each time the cell divides, creating a sort of cellular countdown clock. These images provide long-sought insight into how telomerase works and will help guide the design of drugs that target the enzyme to rewind time—and send the grim reaper packing.

You Are Not the Age You Think You Are

In fact, you have two different ages. One is your chronological age, or the number of years since you were born. The other is biological age, or the time-dependent decline of your body's function and appearance. Your biological age, which represents how quickly the cells in your body are deteriorating, is now considered the true measure of age. It's influenced by your genes, lifestyle, behavior, the environment, and other factors.

There is no one formula for determining your biological age, but scientists are working on it. We do know there's a huge range. That's why some eighty-year-olds have impressive yoga skills and are as active and productive as they were at forty, and other eighty-year-olds have slowed down or succumbed to serious diseases like diabetes, cancer, and dementia. Interestingly, recent studies have suggested that one key variable may be our perceptions.

Data from three longitudinal studies, which tracked more than seventeen thousand middle-aged and elderly participants, showed that most people said they felt about 15 percent younger than their actual chronological age. So a fifty-five-year-old may say they felt about forty-six. However, some felt between eight and thirteen years *older* than their actual age. This so-called subjective age has been implicated in a range of health outcomes. Yannick Stephan at the University of Montpellier, in France, and his team published a study in *Psychosomatic Medicine* on the relationship between subjective age and mortality.[1] It turns out that the group who felt older than their age had an 18 to 25 percent greater risk of death over the study periods, and an even greater chance of getting age-related diseases. So the first step toward immortality is to start thinking you are younger than the birth date stamped in your passport indicates.

Aging Is a Game of Genes

To reduce our biological age, let's start with the source: our genes. The "longevity gene" we discussed earlier is characterized by a specific protein, BPIFB4. An Italian research group, publishing in the *European Heart Journal*,[2] showed how we can introduce this protein into DNA to reverse cardiovascular aging. The study was conducted on mice with fatty plaque deposits on their arterial walls due to a high-fat diet; once given the longevity gene, they were renewed with more youthful vitality. The researchers also experimented on human blood vessels in a laboratory setting. They placed the BPIFB4 protein directly into blood vessels and attained the same rejuvenating results. In the course of their study, the team also found a correlation between healthy blood vessels and higher levels of the protein in the blood. So—who knows?—you may already have the longevity gene.

Gene therapy is changing the conventional wisdom of "you're born with it." While we often think of our genes as fixed, they are actually mutable. In the past, scientists believed that information was passed from parent to child during development, in the womb, setting the stage for many health-related issues. But it turns out that isn't the whole story.

Today, genetics is thought to explain less than 25 percent of the body's main disorders, with the majority caused by environmental, nongenetic factors. A new field of genetic research called epigenetics explores how our environment can influence our genes' expression, opening the door to a new set of health span and life span disrupters (and extenders).

It turns out our genes hold a whole series of possibilities that can be switched on or off by epigenetic triggers such as environment, lifestyle, nutrition, stress, and psychological factors. Today we lost learning how to leverage epigenetic triggers to enhance our health and delay disease by preventing our genes from revolting and mutating. Gene mutations are a permanent alteration in the DNA sequence that makes up your gene.

According to the National Institutes of Health, there are different causes for gene mutation: you may inherit it, which is called the germline mutation, or you may acquire it from your lifestyle and environment impact, called the somatic mutation.

Turning Off the Gene Responsible for Wrinkles and Hair Loss

Remember the study in chapter 01 in which a team at University of Alabama at Birmingham was able to reverse aging-associated skin wrinkles and hair loss in a mouse model caused by a gene mutation just by turning off that mutation to restore the mouse to normal appearance? The instigator was a mutation leading to mitochondrial dysfunction, and, once induced, the mouse develops wrinkled skin and visible hair loss in a matter of weeks. When the mitochondrial function is restored by turning off the gene responsible for it, the mouse returns to smooth skin and thick fur, like that of a healthy mouse of the same age.[3] We can venture to guess that more and more of these powerful solutions will mimic the positive effects of genetic and cellular therapies to slow or halt our aging.

Gene Therapy to Rebuild Knees

GenuCure is a biotech company developing a gene therapy for rejuvenating cartilage. The company has a "cocktail" that will be injected into the knee capsule of people with osteoarthritis, perhaps once or twice a year. Such a treatment could eventually take the place of expensive knee replacement surgeries.

—GenuCureLabs.com

So it turns out that our bad or good behaviors, daily exposures, and other epigenetic factors can trigger—or stop—the ticking of our biological clocks.

A Nobel-winning discovery by the Japanese stem-cell scientist Shinya Yamanaka in 2006 showed how we can wipe clean the epigenetic marks in a cell, giving it a fresh start just by adding four proteins. Called the Yamanaka factor, it is a powerful reprogramming tool for researchers in targeting the mechanisms to stop aging. Even skin cells from centenarians, scientists have found, can be rewound to a more youthful state.

One scientist at the Gene Expression Laboratory at the Salk Institute for Biological Studies in San Diego wants to apply reprogramming not just to cells but to whole animals–and, if they can control it, to human bodies. Juan Carlos Izpisúa Belmonte and his team recently treated an aging mouse with an age-reversal elixir, completely rejuvenating the rodent. In an article for *MIT Technology Review* titled "Has This Scientist Finally Found the Fountain of Youth?" Belmonte explains that there is more work to do and hopes the technology can eventually extend human life span by another thirty to fifty years.

Epigenetic mechanisms may just be the magic bullet that changes the face of medicine. They are already helping to fuel the personalized-medicine revolution in cancer treatment. In the future, epigenetic testing will likely be performed as routinely as blood testing, so that doctors can identify any genetic conditions you might have at different intervals throughout your life.

Dr. Esther M. Sternberg, a professor of medicine at the University of Arizona College of Medicine who is internationally recognized for her discoveries in the science of the mind-body interaction, explained in an interview with Sputnik Futures that chronic stress can change how genes function and may actually speed up chromosomal aging. As you age, chromosomes shorten. Think of chromosomes as shoelaces with little plastic ends on them. As these shoelaces age, the plastic end falls off and they get shorter. Telomeres are like the plastic ends on chromosomes. The effects of chronic stress can make your chromosomes look years older than your biological age.

In their book *The Telomere Effect: A Revolutionary Approach to Living Younger, Healthier, Longer,* Elizabeth Blackburn, a molecular biologist who shared a Nobel Prize for her research on telomeres, and Elissa Epel, director of the Aging, Metabolism, and Emotions Center at the University of California, San Francisco, argue that you can actually lengthen your telomeres—and perhaps your life—by following sound health advice, based on a review of thousands of studies. Some of the findings of their telomere research may contradict conventional wisdom; for example, extreme exercise isn't a requirement for being healthier longer. Nor is a full eight hours of sleep: seven is enough, as long as you feel well rested.

Blackburn first discovered telomeres while at UC Berkeley in the 1970s. But her breakthrough came in 1984 when she discovered telomerase, the so-called immortality enzyme that can add time onto the molecular clock countdown because it is able to

TELOMERES LISTEN TO YOU, they listen to your behaviors, they listen to your state of mind.

—Elizabeth Blackburn, Nobel Prize winner and coauthor of *The Telomere Effect: A Revolutionary Approach to Living Younger, Healthier, Longer*

WE CAN PROVIDE A NEW level of specificity and tell people more precisely with clues emerging from telomere science, what exactly about exercise is related to long telomeres, what exact foods are related to long telomeres, what aspects of sleep are more related to long telomeres.

—Elissa Epel, coauthor of *The Telomere Effect: A Revolutionary Approach to Living Younger, Healthier, Longer*

PEOPLE SUFFERING FROM RECURRENT DEPRESSION or subjected to chronic stress were shown to exhibit shorter telomeres in white blood cells, as documented by a research team at Umeå University in Umeå, Sweden, in 2011, contributing to ongoing research into how depression and chronic stress can accelerate aging.

—*ScienceDaily*[4]

put DNA back onto telomeres. Blackburn and Carol Greider, a molecular biologist, illuminated how we can extend the life span of a cell and were awarded the Nobel Prize in Physiology or Medicine in 2009. Understanding the underlying mechanism of telomerase action offers new avenues toward effective antiaging therapeutics.

Research continues on telomerase and its built-in braking system. "Finding a way to properly release the brakes on the telomerase enzyme has the potential to restore the lost telomere length of adult stem cells and to even reverse cellular aging itself," according to Julian Chen, a professor at the School of Molecular Sciences at Arizona State University.[5] But the trick is in selectively augmenting telomerase activity within adult stem cells. Just as youthful stem cells use telomerase to offset telomere length loss, cancer cells employ telomerase to maintain their destructive growth. Thus, if not done precisely, augmenting and regulating telomerase function can also heighten the risk for cancer development, according to the research published in Medical Xpress.[6]

From the publishers: "Dr. Blackburn and Dr. Elissa Epel's research shows that the length and health of one's telomeres are a biological underpinning of the long-hypothesized mind-body connection. They and other scientists have found that changes we can make to our daily habits can protect our telomeres and increase our health spans (the number of years we remain healthy, active, and disease-free)."[7]

Measuring All Your Aging Variables: The Exposome

Measuring the daily attacks on our minds and bodies, especially the ones that lead to chronic disease, is the key to controlling and treating aging. Globally, approximately one in three of all adults suffer from multiple chronic conditions (MCCs), like diabetes or heart disease. Chronic diseases among people over forty in China are set to double over the next twenty years, owing to changes in dietary patterns, unhealthy behaviors like smoking and excessive alcohol consumption, and pollution associated with urbanization.[8] Nearly 81 percent of US adults age sixty-five and above are likely to experience one or more chronic diseases.[9]

Luckily, there's a new science in town: the exposome. The term encompasses all of the environmental exposures over an individual's lifetime, from conception to death. The exposome was first proposed by cancer epidemiologist Dr. Christopher Wild in 2005, when we didn't have the digital health and environmental sensors we have today, which track everything from UV sunlight to pollution exposure. A key factor in the exposome is the ability to accurately measure the extent and effect of exposures in order to identify the biomarkers that we can regulate.

The best new tool scientists have today in looking for biomarkers is the popularity of at-home DNA testing. Mass genomic profiling companies like 23andMe and Ancestry.com are helping to build a global DNA library that we can borrow from to inform important studies in life extension. Then there are the microbiome sequencing home tests like Viome, where you collect a stool sample at home and return it by mail to get analyzed. If you have done any of these tests, thank you: you have contributed important biological information for every new health innovation in the next century.

The Indoor Exposome

Because humans spend about 90 percent of their time indoors, the built environment exposome merits particular attention. Historically, engineers have focused on controlling chemical and physical contaminants and eradicating microbes; however, we have recently become aware of the role of beneficial microbes. Researchers are seeking ways to selectively control the materials and chemistry of the built environment to positively influence the microbial and chemical components of the indoor exposome.

—*Environmental Science & Technology*[10]

Hair Exposome

Hair exposome is the set of external and internal factors and their interactions that provoke a somatic response related to the hair health. There is industry research underway looking at how pollution and diet affect hair loss; the endogenous and exogenous factors that determine the quality, quantity, and color of the hair, etc. The Commission of Hair Cosmetics of the Beauty Cluster Barcelona (BCB), Centro de Tecnología Capilar (CTC), and Symrise have become instrumental in the development of the concept Hair Exposome (HE), with the purpose of studying hair disorders from a global perspective.

—Skinobs News[11]

Tooth Biomarkers Can Detect an Early-Life Exposome

Tooth biomarkers offer a retrospective, noninvasive tool to directly measure fetal exposures to multiple chemicals. Utilizing the unique aspects of tooth development, the timing of exposures can be determined by growth rings, allowing distinct identification of prenatal (second and third trimester) and postnatal periods. Combined with previous work validating certain metals in teeth, a team of researchers at the Senator Frank R. Lautenberg Environmental Health Sciences Laboratory at the Mount Sinai Health System in New York were able to identify chemical signatures ranging from metals to organic compounds.

—*Environment International*[12]

The Skin Microbiome Tells Where You Live

A study found that differences in the bacterial community structure associated with seven skin sites in seventy-one healthy people over five days showed significant correlations with age, gender, physical skin parameters, and whether participants lived in urban or rural locations in the same state. Adults maintained greater overall microbial diversity than adolescents or the elderly, while the intragroup variation among the elderly and rural populations was significantly greater. Skin-associated bacterial community structure and composition could predict whether a sample came from an urban or a rural resident about five times more accurately than random guessing.

—*PLoS ONE*[13]

Aging May Be Predicted by Your Gut

Some of the data from these DNA and microbiome tests is already being put to great use. Insilico Medicine in Rockville, Maryland, which specializes in pharmaceutical artificial intelligence, is working on a program to predict biological age using several biological markers, including your microbiome. If you haven't yet heard of the microbiome—the tens of trillions of microorganisms that live on and in our body—then mark my words: it will be the key to every personalized health or beauty treatment you can imagine. In fact, the microbiome could someday help physicians treat patients for a medical condition before a single symptom shows up.

Just like a diabetic who checks their sugar levels, we would hope that, in the future, people could check their microbiome and see if they are getting into a state of microbiome shift.

—Julia Segre, PhD, scientist looking at atopic dermatitis flares in children as part of the NIH federally funded Human Microbiome Project (HMP) Source: Healio[14]

Scientists are finding that different organs and fluids in the body each have their own microbial ecosystem, and in 2012, the Human Microbiome

Project was able to sequence the microbiota that normally inhabit the skin, mouth, nose, digestive tract, and vagina. This opened the door for extensive studies about the gut microbiome's role in our health, emotions, weight, immune functions, and brain function, especially the pathways of brain aging that lead to neurodegenerative disorders such as Alzheimer's disease, Parkinson's disease, and depression. Recent findings on the axis of the brain, the immune system, and the gut (known as the "BIG" axis) have begun to identify how a healthy gut microbiome can affect our biological clocks.

Drugs May Soon Address the BIG Axis

The BIG axis is a key focus for creating novel medicines to fight age-related diseases. Science has identified the gut as your "second brain" in connection to the biochemical signaling that takes place between the gastrointestinal tract and the central nervous system and how the gut microbiome is influential in regulating the immune system.

VIOME AT-HOME TEST FOR YOUR gut microbiome uses proprietary technology to identify exactly what microbes are specifically active to help manage weight, digestive issues, mental clarity, skin disorders, and more.

—Viome.com

PURETECH HEALTH IS AN ADVANCED biopharmaceutical company developing novel medicines for dysfunctions of the BIG axis. The company has gained deep insights into the connection between these systems and the resulting role in diseases that have been resistant to established therapeutic approaches.

The Universal Timekeepers

Now that we know most of the factors that drive aging, let's look at the various clocks that can predict our biological age. We'll start with the universal clock: the circadian clock.

The circadian rhythm is a natural, internal process that regulates the sleep-wake cycle and repeats roughly every twenty-four hours. It has been shown to control a variety of physiological and behavioral systems, including metabolism, digestion, cardiovascular activity, endocrine and hormone cycles, sleep cycle, and body temperature, and now it is a focus of aging research. Lab studies from MIT show that a robust circadian rhythm is correlated with a longer life span. Other MIT studies propose that it may soon be possible to prevent and treat the diseases of aging by enhancing circadian function. One area of research expected to expand is the correlation between the use of digital technologies and healthy circadian rhythms.

Hormonal fluctuations are another area of interest. Hormones are natural chemicals produced in one location, released into the bloodstream, then used by other organs and systems. As we age, we produce less of certain hormones, including aldosterone (essential for sodium conservation in the kidney, salivary glands, sweat glands, and colon), calcitonin (acts to reduce blood calcium), human growth hormone (stimulates growth, cell reproduction, and cell regeneration); and, in women, estrogen (responsible for the development and regulation of the female reproductive system).

What Makes Circadian Rhythms Work

- Nearly *every* cell in our body has an internal "clock" that responds to changes in light perception and exposure, helping create our circadian rhythm.
- There's a central clock in the human brain that syncs with our circadian rhythm and regulates body functions such as wakefulness, sleep, body temperature, and hormone regulation according to the time of day. The individual cells of our various organs contain peripheral clocks that take their orders from the central clock. These peripheral clocks help regulate how our body functions, including our metabolism and myriad biochemical pathways.[15]

—*Life Extension* magazine

"Tickle" Therapy Could Help Slow Aging

Scientists at the University of Leeds in London have found that "tickling" the ear with a small electrical current appears to rebalance the autonomic nervous system for people over fifty-five, potentially slowing down one of the effects of aging.

The research found that a short daily therapy, delivered for two weeks, led to both physiological and mental improvements, including a better quality of life, mood, and sleep.

The therapy, called transcutaneous vagus nerve stimulation (tVNS), delivers a small, painless electrical current to the ear, which sends signals to the body's nervous system through the vagus nerve.[16]

The Epigenetic Clocks: Predicting and Reversing Our Biological Time

Only recently have researchers identified several epigenetic clocks that are used for predicting your biological age, measured by various changes to your DNA. These epigenetic clocks can be used to address a host of questions in developmental biology, cancer, and aging research.

The key to these epigenetic clocks is in DNA methylation. DNA methylation occurs by "additions" to your DNA, which often modifies the function of your genes and affects the way your genes express themselves.

Steve Horvath and a research team had been studying DNA methylation at specific genetic locations to see if methylation had anything to do with sexual preference. (It did not.) But, in the process, Steve found the basis for the Horvath clock, a measure of the methylation in some 353 locations in human DNA that acts like a biological age calculator.

Steve was able to develop several clocks, based on readily accessible data about human DNA (upward of eight thousand data sets). This would not have been possible in traditional biological and aging research, where you needed human clinical trials that were both costly and time-consuming, as studies had to span decades to determine any change.

The pan tissue clock is a multi-tissue predictor of age that allows one to estimate the DNA methylation age of most tissues and cell types. In a study Horvath published in 2013, he showed that by using the Yamanaka factor, we can potentially reset the epigenetic age of adult skin cells.[17]

Next came the skin & blood clock (2018), which measures the epigenetic age of skin tissue and also assorted skin cells, like keratinocytes. It applies to the dermis, epidermis, and fibroblast. It can be measured by a punch biopsy from the skin or cultured cells in a dish. Horvath's hope is that this clock could help us identify interventions that slow aging in the skin. There has also been a study to show how the drug rapamycin—originally used to prevent organ rejection and now associated with extending life span by bolstering the immune system—can also slow epigenetic aging in skin cells.[18]

Your GrimAge Predictor

Both the pan tissue clock and the skin & blood clock led to Horvath's DNAm GrimAge clock. In an interview with Sputnik Futures, Horvath explained that the GrimAge clock determines your epigenetic age by analyzing chemical tags that are added to or removed from DNA, which in turn influence which genes are switched on and off. These tags can predict your life span as well as your health span: the time you have left to live disease-free. A tiny drop of blood is all you need to estimate your GrimAge.

Horvath's technology is a major leap forward in proving the connection between DNA methylation and aging, but we still need to better understand the pathways involved: the mechanisms of our molecular DNA clockwork. Horvath believes that we will soon use these epigenetic biomarkers in more advanced studies and human clinical trials.

1:1 with Steve Horvath, the "Clock-Watcher"

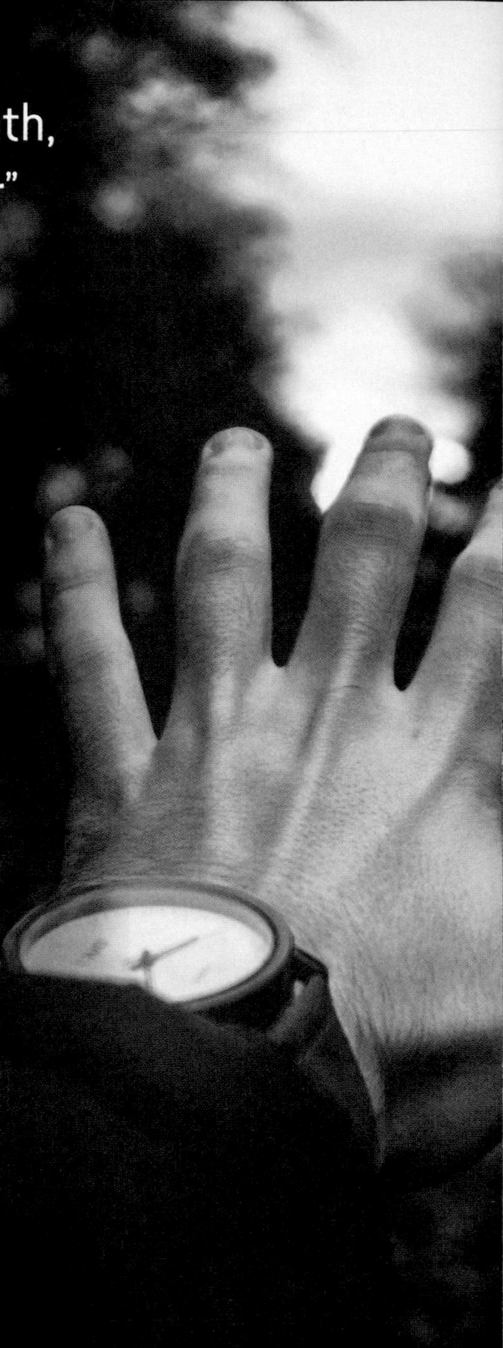

Artists often symbolize mortality by an hourglass, the passage of sand trickling through a small valve. Our twenty-first-century view is that the DNA molecule is the hourglass. Many people have suspected that the DNA is the central molecule of aging, but conclusive evidence was lacking. Epigenetic clocks have provided the strongest evidence that the DNA molecule is the elusive hourglass of mortality.

Around age sixteen, I started to be really interested in questions like: Why do we age? Why is it that some animals live much longer than other animals? I've always felt that it is the central problem of our civilization to greatly extend healthy life span.

Once I received tenure at UCLA, I decided to go back to my true passion, which is to understand aging. And so I turned to the unsolved problem of how to measure aging. I've always felt it's essential to measure aging in order to understand which interventions can change aging, what accelerates aging, what slows aging. I was very fortunate in terms of my timing, because when I started to be interested in that vexing problem, several breakthroughs had taken place.

First of all, there were new ways of measuring molecular changes, the so-called microarrays or chips. And also the scientific culture had changed. People deposited their data sets into public repositories, freely available data, which is different from previous ways of scientific research, where people were hiding their data. I did not need a single dollar of research funding. I simply went to publicly available databases and downloaded data.

I realized that methylation data carry a huge signal for aging in all tissues. As a biostatistician, I was blown away by the strong effects of age on methylation. And I, in essence, dropped everything else. I said to myself, *These epigenetic modifications are the future of aging research. I need to build* biomarkers *based on DNA methylation.* Our UCLA team was the first to build an epigenetic clock based on saliva methylation data in 2011. Two years later I published the so-called pan tissue epigenetic clock. I had a very hard time publishing it because the results were so incredible that nobody believed

METHYLATION, WHAT IS USED IN Horvath's research, is a simple biochemical process: the addition of a single carbon and three hydrogen atoms (called a methyl group) to another molecule. Think of it like tiny on-off switches in your body. Methylation is just one of the several mechanisms thought of as epigenetic: it allows the gears to turn and turns biological switches on and off for a host of systems in the body.

Reveal Your Biological Age Through Epigenetics

Epimorphy, a biotechnology company, offers the myDNAge epigenetic aging clock test based on Dr. Steve Horvath's aging clock, using a urine or blood sample.

—myDNAge.com

A Simple Blood Test Could One Day Reveal Clues About Your Mortality

In a study published in the journal *Nature Communications*, a team of European researchers detail their search for biomarkers in human blood that would let them predict how much longer a person has to live.

Their analysis used blood sample data from twelve groups of similar associations, gathered from a total of 44,168 people between the ages of 18 and 109, After looking at 226 biomarkers in the samples, the researchers determined that fourteen of the blood measurements, including inflammation and fluid balance, were all they needed to predict whether a person was likely to die within the next five to ten years. They found their predictions were right approximately 83 percent of the time—a level of accuracy even they hadn't expected.

—*Nature Communications*[19]

them. But in the end, once the paper came out and the software was available, everybody could easily verify the claims.

There's a whole arsenal of promising antiaging interventions. Given the unspeakable suffering arising from aging, it is imperative that we carefully evaluate promising interventions using epigenetic clocks and additional biomarkers.

A Drug Cocktail Turns Back the Years

In early 2019, scientists at 21st Century Medicine in Fontana, California, conducted the first clinical trial to investigate the possibility that a drug might be able to reverse the biological signs of aging. Nine men, aged fifty-one to sixty-five, took a yearlong drug regimen that appeared to reverse the aging process, leaving them one and a half years younger biologically than when they started. The drug aimed to repair the thymus, a small organ that plays a key role in the immune system and that shrinks with age. The participants were given recombinant human growth hormone (rhGH), because studies suggest it can regenerate the thymus. However, extra rhGH can also trigger diabetes, so the researchers added a supplement called dehydroepiandrosterone (DHEA) and the drug metformin. MRI scans taken at the beginning and end of the trial revealed thymus regeneration, accompanied by improvements in the immune system, in seven of the participants.

The researchers used four different tests of epigenetic age. On average, across the four tests, the volunteers' epigenetic age was one and a half years younger than it was at the beginning of the treatment. This means someone who had an epigenetic age of fifty-five, say, at the beginning of the trial had an epigenetic age of fifty-three and a half at the end of the year-long trial. The most advanced test, GrimAge, showed a two-year decrease in epigenetic versus chronological age that persisted up to six months after the men stopped taking the drug therapy.[20,21]

What Can You Do Today?
Know Your Biomarkers!

So far, we have learned that genes play a part in aging and that lifestyle and other epigenetic triggers are the most critical determinants in our aging timelines. The good news is that, armed with this knowledge, you can start to take control—and perhaps even shut the door on death. The first step is understanding your "biomarkers," indexes in the body that scientists agree quantify aging and tell us how far away disease is.

As we discussed earlier, the pace of aging reaches a sort of tipping point during your fifties, hastening especially for those in poorer health. In an August 2019 article in the *South China Morning Post*, antiaging expert Dr. Lauren Bramley, who owns a medical clinic in Hong Kong, suggests, "If you can get through to fifty without this huge drop off the cliff that so many people experience, then you are kind of laughing because the pace slows down from fifty to sixty and from sixty to seventy."[22]

Because individuals age at different rates, we've long used measures like blood pressure, grip strength, and BMI as indicators of relative health. But today there are many more biomarkers to draw from, especially at the molecular level. We can now get lab tests that measure the oxygen levels of our blood and inflammation induced by sugar, as well as accurate measurements of brain speed and brain capacity, bone density, and genetic markers. In her 2016 TedX Talk, Bramley called on everyone to "know our numbers" just like we know our bank balances and passwords.[23]

BIOMARKER 1: blood sugar. In simple terms, this measures how much sugar you have in your body at any given time, or how prediabetic you are. Sugar sticks to your red blood cells, and too much can make them hard, a condition called glycation. Have your hemoglobin A1c (HbA1c) level checked to see your average blood glucose levels over the past three months. The higher the percentage, the higher your blood glucose levels. Supplements such as metformin can help treat glycation and offset diabetes.

BIOMARKER 2: vitamin D. Studies show that healthy vitamin D levels help prevent cancer, diabetes, obesity, and depression. While we can get a good dose from sunlight alone, fear of skin cancer has taught us to block the rays. The target level for vitamin D is 70 nanograms per milliliter, which you can achieve with a supplement, but different people absorb vitamin D differently, so it is good to check your levels with a blood test before taking anything.

There are a few technical, personalized health centers that can actually help you manage—and, some claim, reverse—your aging and the diseases associated with it. Cofounded by J. Craig Venter, the man who spearheaded the Human Genome Project, Health Nucleus offers a program called CORE, which claims to be the first personalized assessment to reveal a complete picture of your past, present, and future health status. The San Diego, California-based organization is part of Human Longevity, Inc., a genomic-powered clinical research center that uses whole genome sequencing analysis, advanced digital imaging, and innovative machine learning, along with curated personal health information, to deliver each individual profile. Apeiron Center for Human Potential is another clinic working on personalized health and aging, where you can get a deep-dive evaluation of your DNA, biometrics like stress, and 3-D brain mapping. Both Health Nucleus and Apeiron Center's assessments are costly, but they offer membership programs to help you stay on track for healthy aging.

But to keep it simple, according to Bramley, you should know and work on these four biomarkers: blood sugar levels, vitamin D levels, dehydroepiandrosterone (DHEA, an adrenal hormone), and homocysteine. a common amino acid that vitamin B can regulate.

BIOMARKER 3: dehydroepiandrosterone (DHEA). Stress, especially chronic stress, can be a hormone depleter, of the key antiaging hormones in particular. When stressed, the body will convert any available hormone into cortisol, including DHEA. When this happens, your DHEA levels fall, and you become anxious or fearful. DHEA is also responsible for the thickness and wrinkles of our skin and the moisture in our skin and joints. It, too, can be regulated with a supplement if necessary.

BIOMARKER 4: the amino acid homocysteine. Found mostly in meat, high levels of homocysteine are linked to early development of heart disease. It is also associated with low levels of vitamins B_6, B_{12}, and folate. While a blood test may show that your B_{12} level is normal, checking your homocysteine will tell you what's really going on at a cellular level.

—*South China Morning Post*[24]

The Nine Hallmarks of Aging

With the explosion of studies on aging, scientists have identified nine interconnected "hallmarks of aging" determined mainly by our genetics but also modulated by environmental factors. We've covered a few here already, but let's look at the full list.

I. GENOMIC INSTABILITY.

Exposure to smoke, chemicals, or other exogenous agents over time can damage our genome, as can factors like simple DNA replication errors or oxidative stress. Although we have evolved a complex network of DNA repair mechanisms, DNA damage accumulates over the course of our lives, causing mutations in cells that, left unchecked, can lead to cancer.

II. TELOMERE ATTRITION.

Similar to the plastic tips of shoelaces protecting their braided ends (as described earlier), telomeres protect the terminal ends of our chromosomes from deterioration. The normal DNA replication mechanisms in most of our cells are not able to copy the ends of our DNA completely, so the repetitive DNA sequences of the telomere region shorten with each cell division. After a number of replications, this leads to cell growth arrest, limiting the ability of tissues to regenerate as we age.

III. EPIGENETIC ALTERATIONS.

You may wonder how our various tissues and organs can appear so different from one another, since the genetic information encoded in our DNA is exactly the same in all our cells. In fact, DNA is modified with epigenetic information that enhances or suppresses the expression of particular genes as required by different tissue types. For example, if a cell should develop into a liver cell, epigenetic modifications will ensure that the parts of the genome specific to liver cells are expressed, while the parts specific to other cell types are ignored. The aging process often involves changes in our epigenetic code, which can lead to changes in gene expression that affect normal cellular function. In the immune system, for example, this could shift the balance between activating and suppressing immune cells, causing our bodies to be less resilient to pathogens.

IV. LOSS OF PROTEOSTASIS.

In our cells, proteins are constantly being synthesized and degraded in a process known as protein homeostasis, or proteostasis. Proteins are like tools that must be assembled correctly in order to perform crucial cellular functions, and part of the assembly is folding proteins into the proper shapes. Various mechanisms have evolved to stabilize or restore correctly folded proteins and to remove and degrade improperly shaped proteins, which could otherwise accumulate and damage the cell. When these mechanisms become less efficient over time, damaged or aggregated protein components cause dysfunction or even cell toxicity, as seen in diseases like Alzheimer's.

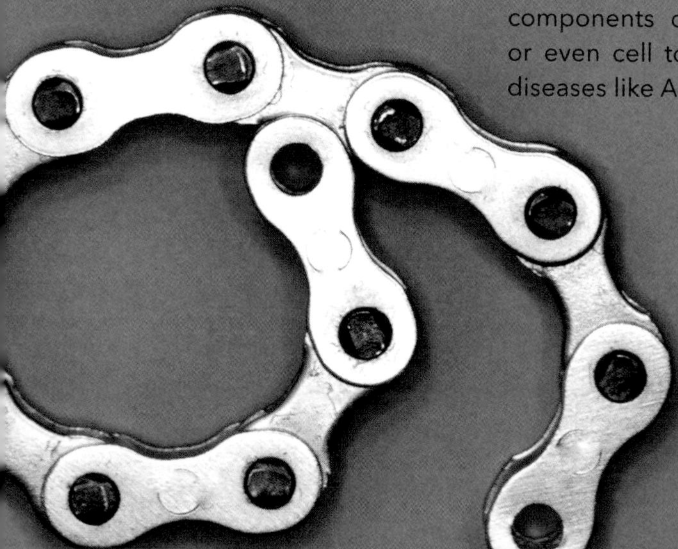

V. DEREGULATED NUTRIENT SENSING.

When nutrients are abundant, cells and tissues respond by storing energy and growing, while nutrient scarcity activates homeostasis and repair mechanisms. In diabetic or obese patients, cells are exposed constantly to abundant nutrients, desensitizing those mechanisms. As we age, the nutrient sensing pathway is further deregulated, and, as a result, cells fail to respond properly to the cues for normal energy production and cell growth.

VI. MITOCHONDRIAL DYSFUNCTION.

Free radicals, or reactive oxygen species (ROS), are a natural byproduct of energy production in the mitochondria. Although they have a role in cellular signaling, in high doses free radicals can damage the cell. The free radical theory of aging proposes that over time, increasing ROS production triggers mitochondrial dysfunction, which causes further ROS production and cellular deterioration. Eventually the cell becomes less efficient at producing energy, and levels of oxidative stress increase, causing damage to other cellular components. As a result, mitochondrial dysfunction contributes to various age-related conditions such as myopathies and neuropathies.

VII. CELLULAR SENESCENCE.

Once cells are subjected to enough stress, DNA damage, and telomere shortening, they enter a state of stable growth arrest called cellular senescence. This is a protective measure to prevent cells with genomic damage from becoming cancerous, but it also prevents old, worn-out tissues from being replenished. Senescent cells change dramatically in their function, most importantly in the molecules they secrete, often proinflammatories that damage the environment of the cells. This leads to chronic tissue inflammation that likely contributes to a variety of geriatric conditions like osteoarthritis and kidney dysfunction.

VIII. STEM CELL EXHAUSTION.

One of the most obvious consequences of growing old is slower recovery from injury, caused by a decline in the ability of our stem cells to replenish damaged tissues. Your stem cells spend most of their time dormant in a niche, but as they are activated to heal wounds and restore tissues, they are also susceptible to telomere shortening, DNA damage, and cellular senescence. Over time this results in stem cell exhaustion.

IX. ALTERED INTERCELLULAR COMMUNICATION.

In order to grow and function normally, our cells must constantly transfer information to each other, secreting signaling molecules to their neighboring cells or even sending molecular messengers through the bloodstream to affect cells and tissues far away. Aging changes not only the signals sent by cells but also the ability of receiving cells to respond. This dysfunctional communication leads to issues like chronic tissue inflammation as well as failure of the immune system to recognize and clear pathogens or dysfunctional cells, increasing susceptibility to infection and cancer.

—Alexandra Bause, PhD, cofounder of Apollo Health Ventures, via Geroscience

Is Nature Immortal?

Or, What Can We Learn from Yeast, Flatworms, and Jellyfish?

According to the scientist David Brin, whether you are a human, an elephant, or a mouse, you begin declining after about half a billion heartbeats.[1] By the age of eighty, we've experienced about 3 billion heartbeats, and soon after, death is inevitable—some even say "natural."

Ummm, someone forgot to clue in the *Turritopsis nutricula*, aka the "immortal" jellyfish.

The natural world is full of creatures and plants with life spans that Dorian Gray could only dream about. The question is: Why do some creatures live longer than mammals, and what can we learn from them? If death is a part of the natural order, why have some species achieved biological immortality?

Biological immortality doesn't mean actual immortality, of course. Even the immortal jellyfish can be squished by an unknowing flip-flop. Rather, it describes organisms that do die but don't seem to age. And they don't contract one of the trifecta of age-related diseases—heart disease, cancer, or dementia—until later in their life. Your pets and most animals experience the onset of age-related diseases roughly 20 percent later in life than humans. So that's a 20 percent longer youth.

Evolution does reward those who find ways to slow the aging process. A paleo diet and a Pilates regiment may help us get to 3 billion heartbeats—but to get to 3 billion and one, we're going to have to learn the secrets of the *Turritopsis nutricula*.

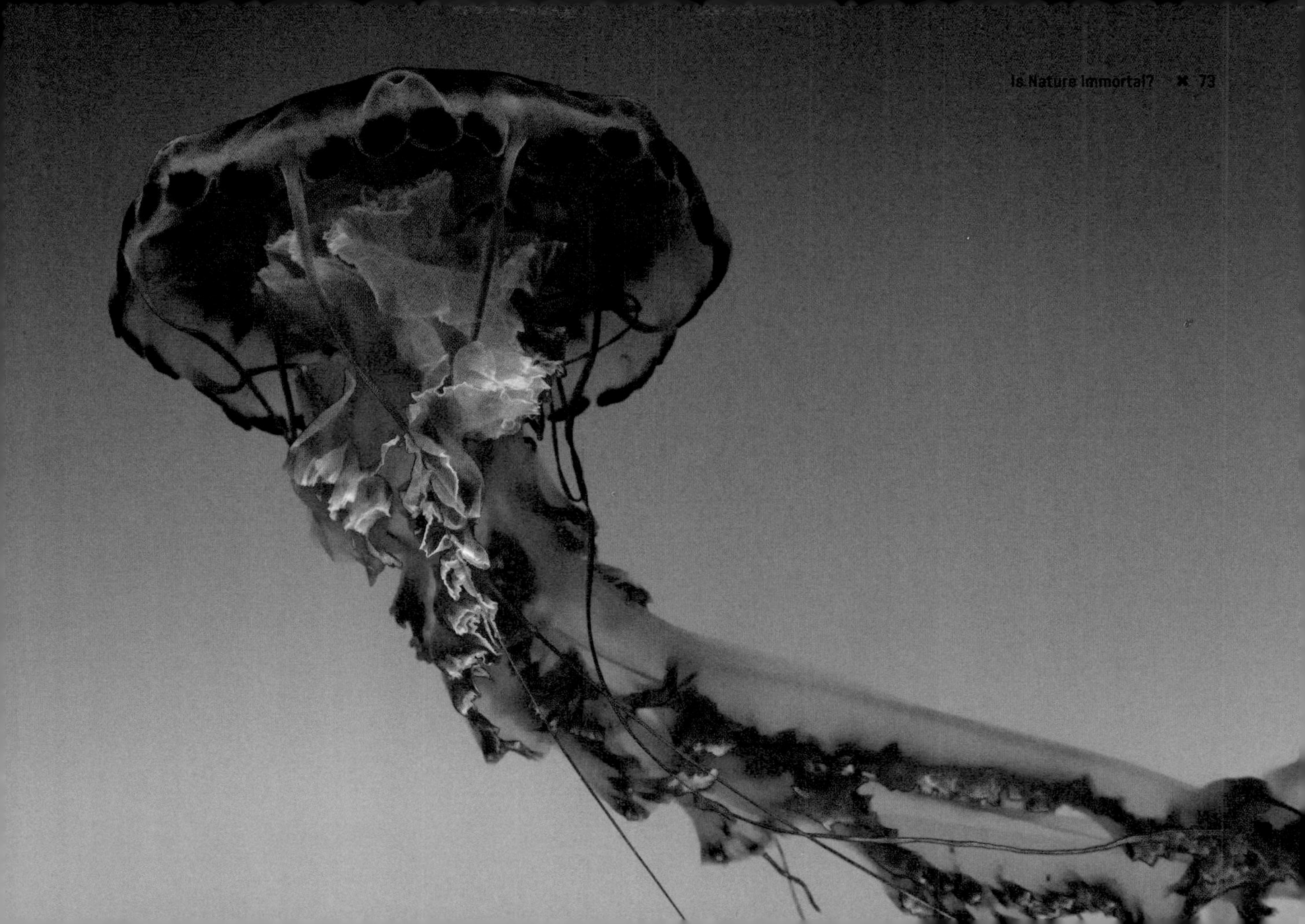

"Ming" the clam, who a team of researchers from Bangor University in Wales claimed to be the oldest animal in the world, was reported to be 507 years old before it died.

Land tortoises can live up to 150 years. Some species of whale can live as long as four hundred to six hundred years. Different types of trees live to be several thousand years old. A little worm known as the *Caenorhabditis elegans* is believed to hold the world record for life span extension at tenfold, or a thousand years.

In an interview with *MIT Technology Review*, Judith Campisi, co-founder of Unity Biotechnology, explained that most of the papers on mice longevity show that their life span can be extended maybe 20 to 30 percent. She explains that humans and mice are "something like 97 percent genetically identical. And yet there's a thirtyfold difference in our life span." Campisi believes that evolution may have had to tweak thousands of genes to help these species evolve to have a thirtyfold greater life span than humans'. But she cautions that "it's unlikely at the present time that we will find a single drug that's going to be able to do what evolution did."[2]

Nature can uncover novel pathways for breakthrough discovery. Microbes, plants, invertebrates, and amphibians have survived for millions of years by developing their own ingenious answers to many of the problems that plague humans as they age. For instance, certain organisms can replace lost or damaged organs and tissues with identical new ones and can regenerate a wide variety of tissues, including spinal cords, limbs, hearts, eyes, and even parts of their brains. Many of these same species have a fascinating skill for repairing and reversing cellular damage, such as cancer. Species with regenerative mechanisms have exhibited the capacity to remodel and integrate cancer back into healthy tissue. Others can age and become young again at a later stage of life—or seem to die and be reborn, like the wood frog, which can stop breathing and whose heart can stop beating for days, even weeks. Unfortunately for humans, the DNA repair mechanisms we do possess fade as we age.

Researchers who study the regeneration and repair mechanisms of our nonhuman friends have found that certain species have a connected network of complex tissues that work together to reprogram and remodel their original structure. According to the life sciences company Bioquark, these capabilities represent a "biological regulatory state reset." It starts with erasing damage in cells, then redirecting the fixed cells into a development program, where they become reintegrated with their cellular neighbors and, finally, reorganized as a community as tissue or an organ.

Humans briefly possess this power of reprogramming and remodeling following fertilization, when egg and sperm first come together to create a new life. But after that it is gone. Bioquark is one of several companies developing novel biologics to reignite these capabilities for a variety of therapeutic applications in humans.

What if Your Body Came with a Reset Button?

The biologic regulatory states of our cells, tissues, and organs represent the central control processes behind our health (as well as our unfortunate transition toward disease, degeneration, and aging). Bioquark's core program focuses on developing a novel class of substances termed combinatorial biologics, which take a unique approach to reversing underlying disease, degeneration, and aging processes, as opposed to only treating the symptoms of such conditions.

—Bioquark.com

So What Can a Jellyfish Teach Us?

The molecule that allows the *Turritopsis nutricula* to turn one type of cell into another is called miRNA, short for microRNA. It is a small, noncoding RNA molecule found in plants, animals, and some viruses. If you recall the mechanisms of epigenetic expression, miRNAs are part of the genetic material that can silence or turn on the regulation of gene expression. Scientists have found that cancer cells have alterations in their miRNA. If we could figure out how to control these miRNAs, you might be able to take cancerous cells and turn them into something else, like muscle, nerve, or skin cells. So the *Turritopsis nutricula* not only is an extraordinary animal that has almost complete control over its own cells but it could also eventually help cure cancer.

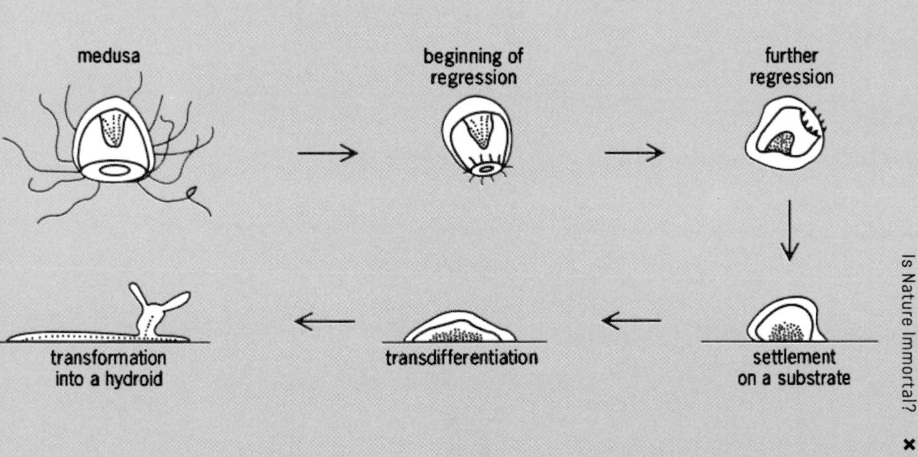

There's Something New to Learn from Every Long-lived Animal

The naked mole rat is incredibly resistant to disease and can live for well over thirty years, a tenfold increase over other rodents. A study published in the journal *Proceedings of the National Academy of Sciences of the United States of America*[3] found that a chemical inside the mole rat's RNA gives its cellular protein factories "the kind of precision that would be the envy of German automotive engineers."[4] That means their cells are less prone to errors as they get older, making mole rats far less susceptible to age-related diseases like cancer and Alzheimer's.

Giant tortoises can live upward of 150 years and are known to have a slow metbolism, allowing them to survive for long periods without food. One Aldabra giant tortoise named Adwaita was around 225 years old when he died. Scientists believe this is due to something in the reptile's genetic code that translates to negligible senescence. If you recall, senescence is the degradation of your cells, turning them into zombies. But senescence in giant tortoises instead slows to a virtual standstill, giving these slow movers a prolonged life.

Lobsters, unlike giant tortoises, grow larger and more fertile with age. Some groups of lobsters produce an impressive amount of telomerase, an enzyme that helps them grow new, larger shells throughout their lifetime, even as they approach fifty or sixty. Unfortunately for lobsters, this ultimately contributes to their demise, as at some point they literally grow too big for their own shells. The energy it takes to molt a new shell becomes too much to bear over time. Lobsters succumb to exhaustion (or shell collapse) at least as often as they do to disease or predators.

Planarians—freshwater flatworms—are the masters of regeneration. They can be decapitated or cut in half and not die; they just make more versions of themselves. Scientists at Nottingham University in the UK managed to create an entire colony of more than twenty thousand wiggly planarians by chopping a single original into several pieces and letting them regenerate. This ability relies on a large number of pluripotent stem cells—the very cells we once thought were "immortal" because of their ability to regenerate into newer cells.[5] Researchers at Germany's Max Planck Institute for Molecular Biomedicine have discovered a protein that is not only required for the maintenance of the stem cell pool in flatworms but may also be active in the pluripotent stem cells of mammals.[6] Despite the rapid progress being made, the technologies for researching the planarian stem cells and their self-preservation mechanisms are still in their infancy.

And then there is the teeny tiny hydra, which carries a remarkably potent set of stem cells in its minuscule body. The hydra is a tiny soft-bodied animal related to the jellyfish that also demonstrates biological immortality. Generally, small animals don't live as long as large ones, but the hydra, measuring just 15 millimeters, is the exception. A biologist claims to have kept an individual hydra alive in the lab for more than four years, and it continued to look young.

The hydra is named after the mythological Hydra of Lerna, which could supposedly regrow its decapitated heads. The hydra's regenerative powers, thanks to its potent stem cells, are crucial during reproduction, as it doesn't usually reproduce sexually: it grows tiny clones of itself. Thomas Bosch, a professor of zoology at Kiel University in Germany, and his colleagues have found that the hydra uses three distinct stem cell populations to replicate all of the various tissues that, taken together, form a fully functioning animal. They identified that all three share one common protein: FOXO (forkhead box O), thought to be a key antiaging protein. In their study published in the *Proceedings of the National Academy of Sciences of the United States*, the researchers outline how the FOXO protein acts as a "hub" in the cell that integrates various molecular signals, including some from the cell's external environment, and regulates stem cell maintenance.[7] Exactly how FOXO prevents the hydra, and in particular its stem cells, from aging isn't yet clear. But, according to Bosch, he and his team are looking at how these environmental signals are integrated with FOXO.

As it turns out, FOXO might work its magic in mammals as well. Studies have consistently revealed FOXO transcription factors as important determinants in aging and longevity.

Monkeys Are Helping Us Understand Human Aging

Give thanks to the marmoset, a small, long-tailed, tree-dwelling monkey that has the same quick mannerisms as a squirrel. Because marmosets are genetically very similar to humans, Professor Corinna Ross, PhD, and her team at Texas Biomedical Research Institute are looking to them for clues to aging gracefully. Marmosets that have been given rapamycin, a drug that's FDA-approved for organ transplants, seem to have improved kidney health and cognitive health. But the researchers also found in their two-year follow-up with the animals that the aging marmosets deemed at risk to die displayed low levels of tryptophan, an amino acid that is linked to the production of serotonin, the chemical in the brain that influences feelings of happiness. They are now questioning whether serotonin can influence aging, since there is already a link between levels of tryptophan and overall health.[8]

Hold the bananas! A shout-out to the rhesus monkeys following a caloric-restriction diet (but still a healthy one) that helped a team of scientists at the University of Wisconsin-Madison's School of Medicine and Public Health discover that caloric-restriction is the key to antiaging in monkeys. Restricting how many calories rhesus monkeys consumed by 25 percent over a span of twenty years made them age differently—and more slowly—than a control group of monkeys that were able to eat whatever they wanted.[9] As we saw earlier, calorie restriction can be effective for humans as well. A two-year-long NIH-supported study found that participants who restricted their calories by 12 percent on average saw decreases in risk factors that contribute to age-related heart disease and diabetes.[10]

Can Yeast Rise to the Longevity Challenge?

Researchers at the University of Southern California think so. They found a mutation in yeast involving an ancient gene called Sch9 that allows yeast to live up to three times its average life span. Not unique to yeast, the Sch9 gene seems to also have a similar function in creatures such as worms, flies, and mammals. Naturally, they wonder if the Sch9 gene is a candidate for gene-based manipulations in humans.

Lead author Valter Longo, an assistant professor and director of the Longevity Institute at the University of Southern California's Leonard Davis School of Gerontology, believes that this gene could have important implications for the treatment of age-related diseases. He and many of his colleagues believe that most people who die as adults die because a particular cell or cell type has lost or modified its function due to aging; that's how we get cancer or Alzheimer's disease or Parkinson's disease. "If we can one day develop drugs that make human cells younger and more resistant to multiple stresses—such as we find in these long-lived organisms—then not only might we extend longevity, we may also protect human cells against a variety of diseases," says Longo.[11]

Not to be overshadowed by its fungi and mammal contemporaries, the phytochemicals found in the plant kingdom have long been heralded as the antioxidant powerhouses to fight aging and help combat a wide array of stressors.

Lycopene is a carotenoid found in red fruits and vegetables that is responsible for their color and is believed to improve human skin texture because it promotes collagen production and reduces the DNA damage that leads to wrinkles. The compounds found in green tea leaves, specifically catechins, are said to help prevent everything from heart disease and cancer to skin aging and weight gain. But the hero antioxidant having the biggest moment is resveratrol, a compound that is commonly found in grapes, nuts, fruits, and red wine, among other things. If you recall, in chapter 01 a few of the supplements and antiaging compounds contained resveratrol, which helps to kick-start cellular sirtuin production to regulate the cell's survival rate.

Fisetin, a compound typically found in your favorite cabernet, is another coloring compound proven to be an effective senolytic in mice studies. A plant polyphenol from the flavonoid group, fisetin can be found in many plants, fruits, and vegetables, such as strawberries, apples, and red grapes. In chapter 01 we learned about senescent or "zombie" cells and how researchers are working on new types of senolytic drugs to target senescent cells while leaving other, "good" cells alone. Researchers working with mice found that fisetin can destroy 25 to 50 percent of senescent cells, depending on the organ and method of measurement.[12] They treated mice toward the end of their life spans with fisetin and saw improvements in health and mobility. These results suggest that we can extend the period of health, or "health span," even toward the end of life. Time to uncork more cabernet? It may take barrelfuls of red wine consumption over two lifetimes or more to get the full benefits of fisetin's zombie-cell-killing power. But, thanks to synthetic biology, there's frontier science in re-creating these life-enhancing natural compounds into effective and sustainable drugs and therapies for antiaging.

The Extremists, and What We Hope to Learn to Survive in Space

Tardigrades, also known as water bears, are microscopic animals that can survive for years without food or water. Thought to be the one species that can outlast all other life on earth, these extremists are now on the moon! Beresheet, an Israeli private lunar spacecraft that crashed on the moon, was carrying the first lunar library, a DVD-sized archive containing 30 million pages of information, human DNA samples, and thousands of tardigrades.

What makes tardigrades famous is a trick called cryptobiosis, in which it brings its metabolic processes nearly to a halt. In this state, it can dehydrate into desiccation, becoming a husk of its former self. But just add water and it springs immediately back to life!

A multinational team of researchers from the University of Edinburgh in Scotland and Keio University in Tokyo have discovered new genome sequences that shed light on both what species tardigrades are—they are now confirmed as a close kin of the roundworm—and the genes that enable their extraordinary ability to survive. According to the study published in the open-access journal *PLoS Biology*, by asking what genes were turned on during the drying process, the scientists identified sets of proteins that appear to replace the water that cells lose, helping to preserve the microscopic structure until water is available again.[13] Professor Mark Blaxter is excited by this, as he thinks "with the DNA blueprint we can now find out how tardigrades resist extremes, and perhaps use their special proteins in biotechnology and medical applications."[14]

Biotechnology will provide the tools to help us mimic the longevity superpowers of nature.

Biotechnology encompasses any process by which nature's existing solutions are bioengineered or biosynthetically re-created into drugs, therapies, and medical devices. Although the term "biotechnology" has only been used for about a century, according to the US National Academies of Sciences, humans have used various forms of biotechnology for millennia.[15]

Synthetic biology is another term that is often used interchangeably with biotechnology, but in fact it refers to a set of concepts, approaches, and tools within biotechnology. The UK Royal Society defines "synthetic biology" as "the design and construction of novel artificial biological pathways, organisms or devices, or the redesign of existing natural biological systems."[16]

Either way, nature-inspired innovation is spawning a new kind of global bioeconomy—competitive, resource efficient, and sustainable. New entrepreneurial forces within it are developing novel foods that are plant-based or culture-derived, finding highly nutritious and more environmentally friendly options beyond meat (in fact, the company Beyond Meat is probably the best-known example here), animal-based dairy, and seafood. The end goal is to create more abundance for all by using fewer resources directly from nature.

The bioeconomy is often linked to the longevity economy, as we are witnessing biology turning into an information science with the ultimate goal of dramatically extending the healthy human life span.

Never in the history of mankind has the gap between imagination and creation been so narrow. With our new and powerful tools of creation including engineered biology, genomics, machine intelligence, robotics, software, and others, we can now create the kind of world we could previously only dream of. Where da Vinci could sketch, today we can build.

–Bryan Johnson, founder, OS FUND

Perhaps the Most Important Breakthrough in Biotechnology Today is CRISPR

If you haven't heard of it, recall *Star Wars: Episode IV—A New Hope*, in which Jedi Master Obi-Wan Kenobi describes the lightsaber as "an elegant weapon for a more civilized age." In our quest to edit the genome, that elegant tool is CRISPR. It allows researchers to easily alter DNA sequences and modify gene function. The protein Cas9 (or "CRISPR-associated protein 9") is an enzyme that acts like a pair of molecular scissors capable of cutting strands of DNA.

The CRISPR-Cas9 system has revolutionized the field of genome engineering, with limitless applications in disease therapeutics, drug discovery, agriculture, biofuels, and much more. For instance, researchers at the Salk Institute have developed a new gene therapy to help decelerate the aging process. Reported in *ScienceDaily*, the findings highlight a novel therapy that can suppress the accelerated aging observed in mice with Hutchinson-Gilford progeria syndrome, a rare genetic disorder that also afflicts humans.[17]

Fruit fly *Drosophila melanogaster* has been used as a model in aging research because it does rapidly age. However, fruit flies with genetic mutations in a gene called *methuselah*, named for a Biblical patriarch who lived to be 969, live significantly longer than normal flies. *Methuselah* is thought to help prevent oxidative damage, and we discussed earlier, its *Indy* (*I'm Not Dead Yet*) gene that proves there is a genetic component to life span. Several other genes in mammals also prolong life span. So the editing potential of CRISPR is understandably exciting. However, editing a single gene in every cell type in the body would be nearly impossible to do. Therefore, we need to

determine which cells can be edited and which oxidative damage genes should be targeted.

Edit Thyself: Will We Become CRISPR Critters?

What would the human race look like without wrinkles? Biotech companies are working on administering CRISPR in a cream that can reach a few centimeters below the skin to boost collagen production and increase elasticity. Several companies are developing topical gels and creams to fight viral infections that live in reservoirs below the skin's surface.

But the most immediate use of CRISPR is to tackle specific diseases we know are associated with age, like Alzheimer's. Researchers at Harvard University found a way to use CRISPR to edit the genome of *APOE4* into a form with a reduced Alzheimer's disease risk.[18] There is plenty more research to come on all fronts: cancer, heart disease, and Parkinson's, just to name a few.

Microbes Will Run the World (Well, Technically, They Already Do)

Microbes—which include bacteria, archaea, fungi, protists, and viruses—have been around for at least 3.5 billion years, and for most of that time they were the only form of life on earth! They are the oldest, most intelligent, most resilient living tools we have in the battle against age.

Bacterial biofilms and slime molds are one way that researchers can observe the astonishing intelligence and beauty of these microbes. Detailed time-lapse microscopy reveals how they sense and explore their surroundings, communicate with their neighbors, and adaptively reshape themselves. Bacteria and similar single-cell forms of life have a surprisingly developed awareness and understanding of themselves and their place within their communities. When they commune in great numbers, they use quorum sensing to communicate directly and pool together their talents to solve problems or take over the environment. *Quanta* magazine explains that those behaviors may be genetically encoded into these cells by billions of years of evolution, but in that sense the bacterial cells are not so different from robots programmed with artificial intelligence to respond in sophisticated ways to their environment.[19]

We spoke briefly about the microbiome, a unique colony of bacteria that controls the health or disease state of any given environment. And we will dive deeper into the weird beauty and uncanny intellect of bacteria in an upcoming book in the Alice in Futureland series. But for one brief example of what bacteria can teach us about immortality, we would like you to meet *Deinococcus radiodurans*, a radiation-resistant bacterium, and *Escherichia coli* (better known by its short form, *E. coli*). These immortal microbes can die and come back to life, thanks to their incredible DNA repair response.

The powers of the *Deinococcus radiodurans* were discovered through a study on radiation. Radiation can kill or damage DNA, later causing cancer and atherosclerosis. Recently at the University of Wisconsin-Madison biochemistry lab, biochemistry professor Michael Cox and his team blasted *E. coli* bacteria with ionizing radiation once a week, causing the bacteria to become radiation-resistant. By the end of the study, these bacteria could withstand one thousand times the radiation dose that kills most humans by dramatically accelerating their DNA repair processes. Understanding this molecular machinery could help us learn how to repair DNA and protect cells in humans and other organisms. For instance, in the distant future, we may see designer microbes capable of making probiotics to aid patients undergoing radiation therapy for some cancers.[20]

THE MICROBIAL HEALTH ECONOMY is just underway and, in terms of value, is projected to reach about $38.7 billion by 2030.

"WE ARE DRIVEN BY OUR GOAL to create a world where chronic illness can truly be a matter of choice and not bad luck."

—Naveen Jain, CEO, Viome

"THERE IS A NEW WAVE of medicines that are coming—ones that are going to be based off of living drugs, living medicines."

—Timothy Lu,
Synthetic Biology Group,
MIT, from an excerpt of the podcast *Live Long* and *Master Aging* with Peter Bowes

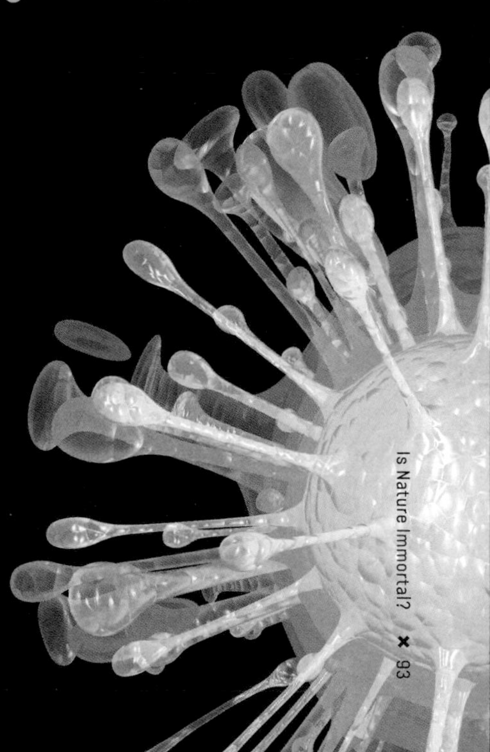

Most drugs are made from natural or synthetic compounds. But biotechnology is enabling us to create "living" drugs that consist of fully functional cells that have been selected and often modified to treat specific diseases, such as cancer. One that holds great promise is CAR T cells, a type of cellular therapy in which immune system T cells are genetically engineered to latch onto and kill a patient's cancer cells. They're being tested primarily in patients with blood cancers such as leukemia and lymphoma.

What if we could screen the body for early signs of disease—from inside the gut? Professor Timothy Lu, a synthetic biologist at MIT, draws on his combined expertise in computer programming, electrical engineering, and microbiology to envision an entirely new approach to preventative medicine, wherein patients swallow tiny capsules packed with diagnostic tools: electronics and genetically engineered living cells. Early signs of cancer, for instance, could be detected without the need for a colonoscopy.

Science continues to find novel ways in which microorganisms and living cells can be our partners in treating disease and enhancing human health. Perhaps, once we fully harness the immortal ingenue of bacteria, we will be able to program our own microbiome for longevity.

In the future, biotech will become domesticated. Visionary physicist Freeman Dyson believes that there will be do-it-yourself kits for gardeners, who will use gene transfer to breed new varieties of roses and orchids, and biotech games for children played with real eggs and seeds rather than with images on a screen. Dyson predicts that once genetic engineering gets into the hands of the general public, it "will give us an explosion of biodiversity."[21]

AMINO LABS WANTS TO MAKE it easy and fun for students to engineer living colors, smells, enzymes, and more by programming bacteria in just a few weeks. The DIY Amino Lab kits engage students with hands-on biological projects that enable them to learn genetic engineering. Amino Labs believes that these are more than just learning experiments; by teaching students from ages eight and up to program bacteria, we will be enabling a new generation of "genetic engineering heroes" that will use biology to help solve the world's biggest problems, from fuel to food to medicine.

—Amino.bio

Field Trip to the Center for PostNatural History

The Center for PostNatural History in Pittsburgh, Pennsylvania, is a contrast to a natural history museum. It is focused on the collection and exposition of organisms that have been intentionally and heritably altered by humans by means such as selective breeding and genetic engineering, a phenomenon the curators refer to as the "postnatural."

—Postnatural.org

Where Does Nature Begin?

It's common and, in many ways, true to say that the line between natural and artificial doesn't really hold anymore. This is another sign of the twenty-first century, the world of the post-human, where we can't depend on an obvious distinction between nature and human culture—because the more we look at nature and begin to manipulate nature, the more we are bringing in human culture. At the same time, we start looking at the patterns of human culture and they begin to look an awful lot, from certain perspectives, like nature. So in many ways this is no longer a viable distinction.

—Erik Davis, author, podcaster, journalist, from his interview with Sputnik Futures

Where Does Nature End?

NEXT NATURE NETWORK, AN INTERNATIONAL network of makers, thinkers, and educators, believes that biology and technology are fusing, and that technology is our next nature. Next Nature Network's philosophy aims to radically shift the notion of what the "nature" we yearn to escape to and debate saving actually is. Through our urge to design our environment, we cause the rising of a next nature that's as unpredictable as ever. It seems that nature changes along with us.

—Nextnature.net

Will One Hundred Be the New Forty?

Anticipating Your Longevity Risk

The real concern is not how much we age but how much we soon may not. With increasing life expectancy, ironically, comes increased "longevity risk."

Do we have the right policies in place to deal with immortality? What quality of life do we want to maintain, and what are the social, political, and financial consequences? Can we design family structures, communities, housing, cities, and insurance policies for mortals that may no longer be mortal? Some fear that boredom may sink in, suggesting that life eternal may be best saved for eternal optimists. To quote Shusaku Arakawa and Madeline Gins, "the artists who refused to die": "How much joy would you be willing to endure in order to secure a possibly uninterrupted future?"

Probably the kid that will be living to 130 is already with us. He has already been born.

—Juan Carlos Izpisúa Belmonte, professor, Gene Expression Laboratory, Salk Institute for Biological Studies, from "The Longevity Issue," *MIT Technology Review*[1]

Without Doing Much, We Are Already Living Longer

Life expectancy has been increasing steadily, and since 1900 the global average life expectancy has more than doubled and is now above seventy years. Experts predict by the end of this century it will easily be over one hundred years. By 2025 we may see five generations working side by side. If a person's life expectancy is around 120, and a new generation is born an average of every twenty-five years, then in 2050 we may see the first eight-generation family.

And it's not just the US. If the pace of increase in life expectancy in developed countries over the past two centuries continues through the twenty-first century, most babies born since 2000 in France, Germany, Italy, the UK, the US, Canada, Japan, and other countries will celebrate their hundredth birthdays.[2]

Radical longevity is a potential multibillion-dollar industry, but it's in everyone's best interest that life extenders are accessible to everyone. There might be some social disruption at first, but life extenders will eventually trickle down the socioeconomic classes, just like any other piece of technology.

Today we associate old age with infirmity and disability. Nobody wants to be old because nobody wants to be sick, but we forget that science will prolong good health too. Put it into perspective: if the length of our lives is doubled, so, too, will be the best years of our lives.

This potential new "fourth stage" of life, where we are vital and productive citizens well into our eighties, nineties, and even hundreds is not just a scientific challenge but also a socioeconomic one. With increased health span, we may also see people having more children even later in life. Which begs the question: Are we as a family, a society, and a nation prepared for the financial burden of more people living longer lives? Extreme longevity will lead to extreme generations, and unprecedented financial, health, social welfare, and environmental stress on both developed and developing countries. To maintain and even increase socioeconomic prosperity for a nation experiencing exponential growth in population, especially among the aging, governments need to encourage a larger working-age population.

However, the way we currently define "working age population" will also be challenged, as we will have healthier and more fit people who are in their eighties and still capable of working, especially as the gig economy expands. Then you have the threat of artificial intelligence taking over some of the tasks and roles that were once relegated to an experienced human. There are many optimists and experts who predict that both AI and humans will be working side by side to create once-unimaginable discoveries and innovations, to the benefit of everyone. And why not? Society will have a wealth of knowledge and experience from a pool of seasoned individuals who can help train AI. (More on the collaborative potential of humans and AI in an upcoming Alice in Futureland book.)

What we have brewing today is a perfect storm of demographics, job availability, health care access, and economic stability. "The Baby Bonanza" was a seminal article published in the *Economist* in 2009, exploring Africa's demographics at a time when global powers were looking at

Africa as the next sleeping giant.[3] It referenced the "demographic dividend," a term coined by Harvard economist David Bloom, who emphasized the consequences of population change on economics. The demographic dividend posits that a decline in a country's birth and death rates, and a larger working-age population, can enable accelerated economic growth in a country. But there is a catch: with fewer births each year, a country's young population declines in relation to the working-age population, which can lead to economic stress. Similar dynamics will occur when you have a much larger older population living longer. In developed nations like the US and most of northern Europe, birth rates are in decline. In developing countries like Africa, birth rates continue to rise, but there is not adequate healthcare for all. Most governments in developing markets are not equipped to handle the increased aging populations expected. And of course we all need to ask ourselves: Are we personally equipped not only for retirement but for our likely longer life spans and health spans?

Aging Projections from the United Nations:

In 2045, the number of people sixty or older will be higher than the number of children worldwide for the first time in history.

By 2050, the number of older people will double to 22 percent of the total population, with the most rapid increase occurring in developing countries.

Among older people, the fastest growing group is people age eighty or over. Today, about one in seven older adults is over eighty.

—United Nations.org

Rapid Urbanization and Aging Population Will Create a Sunrise Industry

Senior housing and care will become a business, not just a government service. The world urbanization prospects by United Nations Department of Economic and Social Affairs' population division notes that the largest urban growth will take place in China, India, and Nigeria. These three countries will account for 37 percent of the projected growth of the world's urban population between 2014 and 2050.

—United Nations.org

Who Will Finance "Forever"?

Now that we can measure our biological age, based on the length of our telomeres or the biomarkers in our cells and tissues, you may be soon find out that your sixty-five-year-old self has the body of a forty-five-year-old. It should be exciting, but it does pose challenges in your financial planning. According to Professor Moshe Milevsky at York University in Toronto, today's foremost authority on retirement finance and author of *Longevity Insurance for a Biological Age: Why Your Retirement Plan Shouldn't Be Based on the Number of Times You Circled the Sun*, chronological age is a poor metric that we should not be using. Thinking in chronological years distorts your work strategy, asset allocation, life insurance, retirement planning, and everything else. Milevsky advocates a new individual financial strategy, such as a greater proportion of equities or an annuity. His book examines the personal financial implications of extreme longevity and how to guarantee a sustainable income stream for the remainder of your biological

insurance today, because tomorrow you might find out you're younger."[4]

Milevsky also predicts that longevity will have a dramatic impact on public retirement policy as well, raising the bar on the age you can claim, and start collecting on, government pensions or IRAs. Shifting our age measures to biological ones will help make longevity risk as salient as mortality risk, which will one day make annuities as "legitimate" as life in-

In an article in *Barron's*, Milevsky acknowledges that "there are going to be winners and losers using biological age."[5] Consider that the world's poorest have a tendency to live shorter life spans due to unhealthy diets and lifestyles, along with limited access to health care and preventive therapies. According to Milevsky, these people, or those who discover that they're actually biologically older than their birth date would predict, shouldn't have to wait until the current designated age (such as seventy) to receive their pensions. On the other hand, the biologically fit may have to wait longer. It will be a thorny problem for governments and private business alike.

Can Life-Extending Treatments Be Made Available to All?

Is there a way to increase the general welfare without creating drastic inequities in health span? These are some of the financial and ethical questions we will be addressing as the science gains momentum. Ethicists such as Colin Farrelly at Queen's University in Kingston, Ontario, and Tom Mackey at the Georgetown University Law Center have argued for the widespread provision of life-extending drugs by public health services. In an article on Aeon.co, they claim this would result in a more equal society. They reason that it will make more sense economically to prioritize health care that combats age-related illnesses (discussed in chapter 01), as it will free up the stress of having expensive diseases to treat for a large part of an individual's life in the absence of preventive measures.[6] In other words, it will be cheaper to prevent or slow down aging than to treat the diseases themselves.

Of course, there are still societies, including the US, that are fighting to make health a human right in the first place. Health priorities differ between societies, religious denominations, and countries, and this will add to the complexity of who will have access to life-extending treatments and at what cost.

What Is Your Longevity Risk?

In the future, there may be an algorithm that will tell us when, financially, it is the right time to die. It may be as easy as a metric on your phone that measures and reports how old you really are and projects how much you will likely be spending in retirement.

We will also see new types of age-appropriate products either to address a vitally healthy older age or to help increase health span. Financial interests may have far more influence on the development and application of biomedical research, consumer products, and services than scientific evidence ever can.

THE REALAGE TEST is an online longevity calculator, part of Sharecare, a health-centric online community, and affiliated with TV personality Dr. Mehmet Oz. Over 43 million people have taken the four-part test, which takes about twenty minutes to complete. The company claims you will get better accuracy by knowing such things as your cholesterol, blood pressure, and any vitamins you take, including the doses.

—Sharecare.com

WHAT CAN I DO NOW to reverse aging? This festival wants to give you the answers. RAADfest is an initiative of the Coalition for Radical Life Extension, a not-for-profit organization reaching out to groups and individuals who already have an interest in radical life extension and physical immortality. An annual event, RAADfest is part festival and part antiaging seminar, with a marketplace of products and companies you can purchase from. It brings together experts on the topic of reversing aging, new advances, and practical applications to support your goals to reverse aging now and for the future.

—RAADfest.com

THE METHUSELAH FOUNDATION, a biomedical charity cofounded by David Gobel and Dr. Aubrey de Grey in 2001, has a mission to make ninety the new fifty by 2030. Just like the gene we talked about earlier, the foundation was named after Methuselah, the grandfather of Noah in the Hebrew Bible, whose life span was recorded as 969 years. The foundation has awarded more than $4 million to support research and development into regenerative medicine that could extend life spans. Its Methuselah Mouse Prize encourages scientists to work on antiaging research and treat aging as a "medical condition."

—Mfoundation.org

And Now You Can Even Meet Your Future Self

Walter Greenleaf, a behavioral neuroscientist and a medical technology developer at Stanford University, has been working on software that uses information from wearables and other sources about how you are treating your body to show how you will age through a computer-generated 3-D avatar. In an interview with Sputnik Futures, he shared some interesting results from a virtual reality (VR) study with students at Stanford University. It turns out that seeing an image of themselves age as few as five years made the students highly motivated to take better care of themselves in the present.

As a medical researcher, Dr. Greenleaf's focus has been on age-related changes in cognition, mood, and behavior. He has been working in the field of medical virtual reality for over thirty years and currently works with emerging technology start-ups that are applying VR and augmented reality (AR) technology to health care, and the research labs studying the effectiveness of its application. He and the team at Stanford's Virtual Human Interaction Lab pioneered the use of VR to show people their future selves and had some impressive results, such as behavior change and motivation to save money. And no blood test required!

Dr. Greenleaf, behavioral neuroscientist, on how the future self works and the value of this AR concept for health care and the individual:

We can take a picture of a person, also record their voice, and, using AI technology and computer graphic technology, we can create an avatar that represents their future self. The avatar looks like them; it sounds like them; it is a realistic representation of how they will appear and sound twenty years from now, thirty years from now. In other words, it is their "future self." By using voice recognition and AI technology, you can have a dialogue with that future yourself.

What's the value of that? Well, right now it's often hard for people to appreciate how the choices they make on a day-to-day basis are going to affect their future. You know, it's hard to manage our weight well; we don't see the effect of eating something that might have too many calories until weeks later or months later.

In a similar manner, it's difficult to see the effect of not taking medication if we really need to take it but we're not in the mood to take it. The decisions we make regarding taking our medications certainly affect our long-term health; yet, when you look at the statistics, in general, people have trouble adhering to a treatment plan—even if it can have dramatic negative consequences to not do what is necessary.

For people who have trouble with substance abuse, or modulating their use of alcohol or nicotine, we can show them their future self and they can see how their choices and behavior affects their body physically. We can also have their future self have a talk with them and say, "Hey, what are you doing? Here's the house [you're] living in now and it's looking pretty good, [you're] looking pretty happy, but here's the other house and here's how [you're] going to look if you don't change your behavior." It's a powerful technique, and it does work.

We've had success at motivating people to exercise, to take their medications, and to do the difficult work required to recover

Medical Avatar Is Like a Future Selfie

Medical Avatar helps you shape better personalized mental models by visualizing possible future versions of yourself. Using photos and self-tracking data, the app generates 3-D models of your body over time. Its algorithms take past data to calculate future probabilities, such as what you will look and feel like in a year—or five years or ten years—if you eat a healthier diet, sleep better, and exercise more.

—Medicalavatar.com

from an injury or to address addiction. Here is an example that I like to cite: a study conducted by Jeremy Bailenson and his team at Stanford University. They gave cash to students who were in a study group; they told the students that they could do whatever they wanted to with the cash. The students who got to meet and interact with their future self clicked on the link that allowed them to put some of the money aside for retirement. Those who did not meet their future self were not inclined to do that. When you can see how your actions now have a direct effect on your future, it can motivate you to make the right choices. It is very difficult to accomplish this without some form of feedback about consequences. We can use VR and AR technology to provide feedback in a dynamic and compelling manner.

The Key to Becoming a Centenarian: Check Your Personality

The New England Centenarian Study has enrolled more than 1,500 centenarians from around the world in the past fifteen years. When the centenarian study began in 1994, the demographic percentage prevalence rate of centenarians was one per ten thousand, making centenarians one of the fastest-growing—if not *the* fastest-growing—segments of the population. In 2010 there were about seventy thousand centenarians in the United States; 85 percent of centenarians in 2010 were women, and 15 percent were men. In a 2009 article titled "100 Is the New 65" in *Greater Good Magazine*—notice how we've moved the bar lower, to age forty, due to all the discoveries this past decade?—the study's director, Thomas Perls, claimed that these participants dispel the belief that the older someone gets, the sicker he or she becomes. Instead, the study found that one's age correlates to how healthy one has been.[7]

For years, we have known that to reach extreme old age correlates basically to what your doctor and public health officials advocate: Don't smoke. Drink in moderation. Eat healthy. Exercise regularly. It's easy to know what it takes to stay healthy. It is more difficult to believe that we have the social and emotional power to drive maximum life span and health span.

There is growing evidence that our personalities can affect our longevity. The Georgia Centenarian Study, which ran from 1988 to 2006, identified a cluster of personality traits common among the centenarians involved. One trait they displayed in relatively high numbers is what psychologists label "competence"–the ability to achieve goals–and "conscientiousness," or self-discipline.[8] Just as mindfulness apps are teaching us to lower our stress, perhaps augmented reality technologies like seeing our future selves will instill in us a new sense of self-discipline we can practice now and watch ourselves grow younger every day.

The Rise of the Supercentenarians

Th[illegible] extension is bound to have a profou[illegible]mily structure and intrafamily dynamics. Supercentenarians, people who are 110-plus years old, occur today at a rate of about one per 5 million. In 2010 there were sixty to seventy supercentenarians in the US. In 2017 the New England Centenarian Study enrolled its 150th supercentenarian. With women making up 85 percent of the centenarian cohort, the study suggests that the female prevalence among supercentenarians may increase to about 90 percent.[9]

Dave Asprey, founder of Bulletproof, claimed in an *Entrepreneur* article that he plans to live to 180.[10] He is among hundreds of biohackers who follow a strict regimen to increase energy and brainpower and extend their longevity. He created the Bulletproof coffee phenomenon a few years ago, urging health enthusiasts to add two tablespoons of butter to their coffee to stimulate weight loss and spark brainpower throughout the day, and claimed the key to longevity is tapping into the mitochondria, which are known as "powerhouses of the cell." Asprey's techniques for maximizing his mitochondria to work at their maximum capacity include taking various supplements, taking cold showers, practicing intermittent fasting, and getting better sleep.

Thirteen Antiaging Supplements to Turn You into Benjamin Button

It's no secret that I plan to live to 180, and my antiaging supplement stack is one of the ways I'm going to make that happen. Too many people think that aging involves forgetting your name and leaving your car keys in the fridge. It doesn't have to be that way. Specific antiaging supplements help you turn back the clock so you can keep kicking butt and taking names for years to come.

—Dave Asprey, founder of Bulletproof, via Bulletproof.com

- Nicotinamide riboside
- Fisetin
- Collagen protein
- Acetyl-L-carnitine
- Whey protein
- Pyrroloquinoline quinone (PQQ)
- Polyphenols
- KetoPrime
- D-ribose
- Zinc and copper
- Vitamins A, D, and K
- Curcumine
- Apigenin

—Bulletproof.com

Where You Live May Impact Your Chances of Being a Centenarian

A few places in the world are called "blue zones." The term refers to geographic areas in which people have low rates of chronic disease and longer life spans. Dan Buettner, a National Geographic Fellow, discovered five blue zones in the world where people live the longest and are healthiest: Okinawa, Japan; Sardinia, Italy; the Nicoya Peninsula, Costa Rica; Ikaria, Greece; and Loma Linda, California (see sidebar).

Buettner's identification of these longevity hot spots was built upon the demographic work by Gianni Pes and Michel Poulain, first outlined in the journal *Experimental Gerontology* in 2004, which identified Sardinia as having the highest concentration of male centenarians in the world.[11] Their collective research also found that all blue zone areas share nine specific lifestyle habits that they call the Power 9, which is outlined in Buettner's books *The Blue Zones: Lessons for Living Longer from the People Who've Lived the Longest* and *The Blue Zones Solution: Eating and Living Like the World's Healthiest People*. (See sidebar.) Buettner has since launched the "Blue Zone Project" to help people live longer through community transformation programs that lower health care costs and improve productivity. Results from cities, businesses, and communities in Texas, Iowa, Minnesota, and California that adopted the Blue Zone Project goals have been remarkable, including double-digit drops in obesity, smoking, and body mass index (BMI).

Do You Live in a Blue Zone?

Barbagia region of Sardinia: mountainous highlands of inner Sardinia with the world's highest concentration of male centenarians.

Ikaria, Greece: Aegean island with one of the world's lowest rates of middle-age mortality and the lowest rates of dementia.

Nicoya Peninsula, Costa Rica: world's lowest rates of middle age mortality; second highest concentration of male centenarians.

Loma Linda, California: home to the world's highest concentration of Seventh-Day Adventists. They live ten years longer than the average North American.

Okinawa, Japan: females over seventy are the longest-lived population in the world.

—BlueZones.com

The Power 9 Evidence-Based Common Lifestyle Denominators among All Blue Zones

1. Move naturally.

2. Feel a sense of purpose (worth up to seven years of extra life expectancy!).

3. Shed stress.

4. Stop eating when your stomach is 80 percent full.

5. Eat a plant-based diet.

6. Wine at 5 p.m.: one to two glasses with food (and friends) daily.

7. Belong to some faith-based organization.

8. Put loved ones first.

9. Pick the right tribe to support healthy behaviors. (Okinawans created the moai, a circle of friends committed to each other for life.)

—BlueZones.com

Nuchi Gusui (Food Is Medicine)

The Kuakini Honolulu Heart Program is one of the largest, longest, and most comprehensive medical studies of centenarian men that seeks to crack the secrets of long life and healthy senescence.

Dr. Bradley Willcox, a professor of geriatrics at the University of Hawai'i's medical school and a researcher with the Kuakini Health System, and his twin brother, D. Craig Willcox, a geriatrics and gerontology investigator at Pacific Health Research & Education Institute, traveled to Okinawa in the 1990s and teamed up with Dr. Makoto Suzuki. Together, they wrote two *New York Times* bestsellers on the Okinawan diet and lifestyle, which includes daily activities like gardening that ground one in nature, meditation, and eating healthier while consuming less. The Okinawans have a simple diet consisting mainly of yellow, orange, and green vegetables, soy, and legumes. They favor purple sweet potatoes over white rice, and meat (including pork), dairy, and seafood are eaten in small amounts. The entire diet is low in sugar and grains: Okinawans consume about 30 percent less sugar and 15 percent fewer grains than mainland Japanese.

If Americans lived more like the Okinawans, "80 percent of the nation's coronary care units, one-third of the cancer wards, and a lot of the nursing homes would be shut down."

—"The Okinawa Way," *Newsweek*, December 12, 2003

FROM THE WILLCOXES AND SUZUKI'S book *The Okinawa Program: How the World's Longest-Lived People Achieve Everlasting Health—and How You Can Too*:

THE OKINAWA PROGRAM AND THE Okinawa Diet Plan are based on the landmark scientifically documented twenty-five-year Okinawa Centenarian Study, a Japanese Ministry of Health–sponsored study.[12] The book reveals the diet, exercise, and lifestyle practices that make the Okinawans the healthiest and longest-lived population in the world and includes an easy-to-follow "Four-Week Turnaround Plan" with nearly one hundred fast, delicious recipes and a moderate exercise plan.

Okinawa, Japan, and a Gene Called *FOXO3*

Today, much of Willcox's research revolves around a gene called *FOXO3*, which makes you more metabolically efficient. We all have the *FOXO3* gene, but only some of us have what Willcox calls the "longevity version." If you have the *FOXO3* GG genotype, rather than the more common *FOXO3* TT genotype, you have almost three times the chance of living a long life. In the future, it may be technologically possible to insert the *FOXO3* gene variant into humans, but the first approach should be to upregulate your own system naturally through lifestyle and diet.

The Centenarians in Okinawa Are Good Luck Charms

There's a cultural habit in Okinawa called *ayakaru*, where the ninety-seventh birthday is very special. It's a major goal of a lot of people, that they want to make their ninety-seventh birthday. And they celebrate the ninety-seventh with a ceremony called *kajimaya*, where they are paraded around the village in the back of an open truck and they are carrying a pinwheel—and that pinwheel celebrates or symbolizes their return to youth, and they are given a bit more leeway after ninety-seven to act more like children. People believe if they touch this person, they can achieve the power of healthy longevity.

—Dr. Bradley Willcox, internist and gerontologist in the Department of Geriatric Medicine, John A. Burns School of Medicine, University of Hawai'i, and a researcher at the Kuakini Medical Center; interview with Sputnik Futures

Paolo Antonelli, director of research and development and senior curator of architecture and design at the Museum of Modern Art (MoMA), posed the question "Is old age a new form of youth?" at the MoMA R&D Salon 22 on New Aging. This is the same burning question challenging architects and urban planners who have been developing communities for the "third age" population—people who have retired yet are in relatively good health. In the US alone, the retirement-community industry generated over $70 billion in revenue in 2018, according to IBISWorld research. But visionary architects are looking beyond recreation for retirement living to improving longevity with built environments that challenge the senses.

Designing for Immortality

Shusaku Arakawa and Madeline Gins were visual artists, conceptual writers, and self-taught architects who believed that by radically recalibrating the way you move through a space or built environment, "humans could solve the ultimate design flaw: death."[13]

The husband-and-wife team developed a design ethos called Reversible Destiny, which aims to promote longevity by stimulating the body and mind. As architects, they wanted to create spaces that challenged the body, thereby fighting deterioration. In their lifetimes, Arakawa and Gins completed five projects—three in Japan, two in America—that are like playgrounds for reorienting space and reimagining movement. According to their manifesto, *Architectural Body*, they did not accept death as a limit on the human condition. Mortality is not only negotiable, they believed, but reversible.[14]

The Reversible Destiny Lofts MITAKA—in Memory of Helen Keller—were their first residential works, a complex of apartments

REVERSIBLE DESTINY

death, not the word
but the event, becomes obsolete

nondeath without end

denecessitates dying

dying becomes extinct

no more irretrievable disappearances

vintage nondying

ongoing regeneration

no destiny but a reversible one–
the pair as inseparable: reversibledestiny

its motto: death is old-fashioned

Body Proper + Architectural Surround = ARCHITECTURAL BODY

built in 2006 in Mitaka, a suburb of Tokyo. Arakawa and Gins used what they called the philosophy of "procedural architecture" to challenge and stimulate the senses. The nine residential units utilize just three shapes (the cube, the sphere, and the tube) arranged in stacked forms. Each apartment has a circular room with a kitchen at its center, and the floor of the central space is deliberately uneven, with vertical poles to assist in moving through it. The off-center design plays with our sense of proprioception, one of the more than twenty-one senses we use to experience the world around us. Also referred to as kinesthesia, it is the sense of self-

movement and body position. Navigating a Reversible Destiny apartment would force you to move more unnaturally, perhaps increasing the effects of daily movement, as outlined in the Blue Zones lifestyle practices.

ST Luk, project manager at Arakawa and Gins's New York–based Reversible Destiny Foundation, explained to Sputnik Futures in an email that "They thought about death as a process, which the body is constantly trying to fight against." Arakawa and Gins believed that once humans adapt to whatever space we are given, and that as soon as you feel comfortable, your body begins to deteriorate. So their architecture is meant to keep you from fully adapting.

Most people, in choosing a new home, look for comfort: a serene atmosphere, smooth walls and floors, a logical layout. Nonsense. People, particularly old people, shouldn't relax and sit back to help them decline. They should be in an environment that stimulates their senses and invigorates their lives.

–Shusaku Arakawa, visual artist, conceptual writer, architect

How to Experience Reversible Destiny

As you step into this unit, fully believe you are walking into your own immune system.

Treat each room as if it were a direct extension of you.

—Two of the twenty-two instructions belonging to the Directions for Use of the Reversible Destiny Lofts

—RDLoftsMitaka.com

Welcome to Yoro Park, Where Play Is for Adults!

Set in one corner of Yoro Park, famous for its waterfall, is a Reversible Destiny-themed park designed by Shusaku Arakawa and Madeline Gins. "The Site of Reversible Destiny—Yoro" spreads across approximately eighteen thousand square meters and consists of a main pavilion called the Critical Resemblance House and the vast, bowl-like Elliptical Field. The site's scale and various mysterious objects surprise and delight visitors of all ages. Adults revert to childhood feelings and are inspired to jump and discover various possibilities of using and balancing the body.

Perhaps to become well-rounded 120-year-olds, we should consider the Fun Theory Sequence by artificial intelligence theorist Eliezer Yudkowsky. Fun Theory is the field of knowledge that deals in questions such as "How much fun is there in the universe?" and "Will we ever run out of fun?" and "Are we having fun yet?" and "Could we be having more fun?"[15]

A key insight of Fun Theory is that eudaimonia—the theory that happiness is the ultimate human goal—is complicated, as there are many properties which contribute to a life worth living. Humans require many things to experience a fulfilled life: aesthetic stimulation, pleasure, love, social interaction, learning, a sense of security, and much more. If we have the chance to live longer, healthier lives, should happiness be part of the equation? And how do we value it? This is part of the challenge facing the digital immortalists today. Perhaps the work you need to do today to become a supercentenarian (why not?) is to have more fun and to think of how you're going to curate your future self as you prepare to log every memory, every bit of advice, into a digital record of your time on earth for your future offspring and friends to interact with. Let's practice: Smile. Copy. Upload. Feel good?

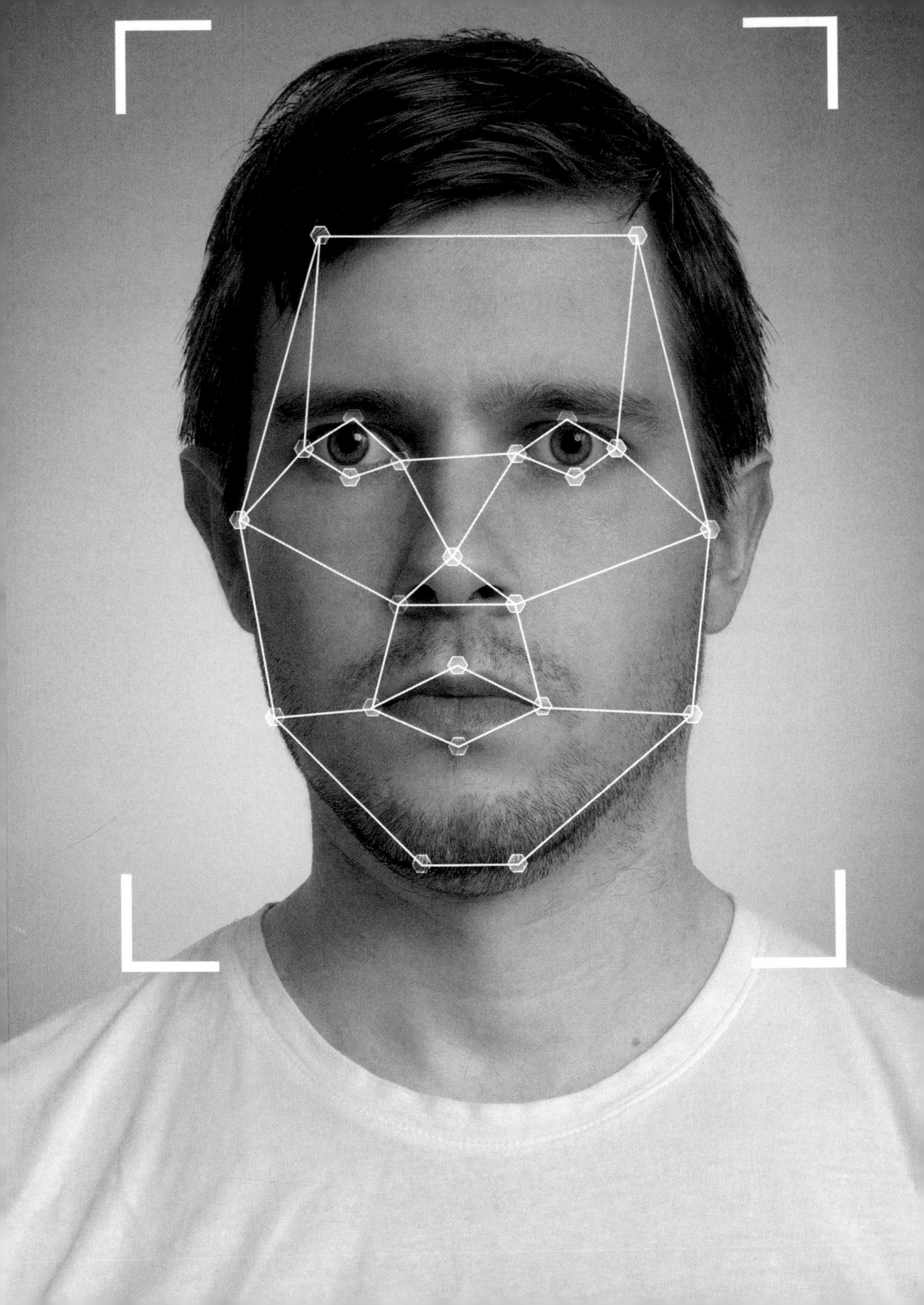

Will We Become Software?

Upload, Download, Augment, and Upgrade Yourself

We are not software just yet, but it might be a good idea to plan ahead while we still have a meat brain. After all, the cyborgs are coming.

And even now our digital trails live forever as we curate ourselves in cyber-eternity. From reverse engineering the brain to implanted neurobots to a distributed machine-intelligence network that can map an individual's every thought, action, and mood into an app that can be accessed for generations to come, we are reframing what it means to be mortal.

As the pace of robotics and AI accelerates, we need to come to grips with the new intelligence we are creating. To quote science fiction writer Vernor Vinge, who coined the term "singularity," in the relatively near future we will see change as great as the rise in humankind within the animal kingdom. He continues to explain in an interview with Sputnik Futures that he believes it is very likely that in the next ten to twenty years there will be people much smarter than we are, a greater difference than now exists between a person and a goldfish. Enhancing intelligence, according to Vinge, will at some point lead to a positive feedback loop; more intelligent systems can design even more intelligent systems, and much quicker than the original human designers. This positive feedback will be powerful enough that within a very short time—months, days, or even hours—the world will be transformed beyond recognition and suddenly inhabited by superintelligent beings. Goldfish, are you ready for your upgrade?

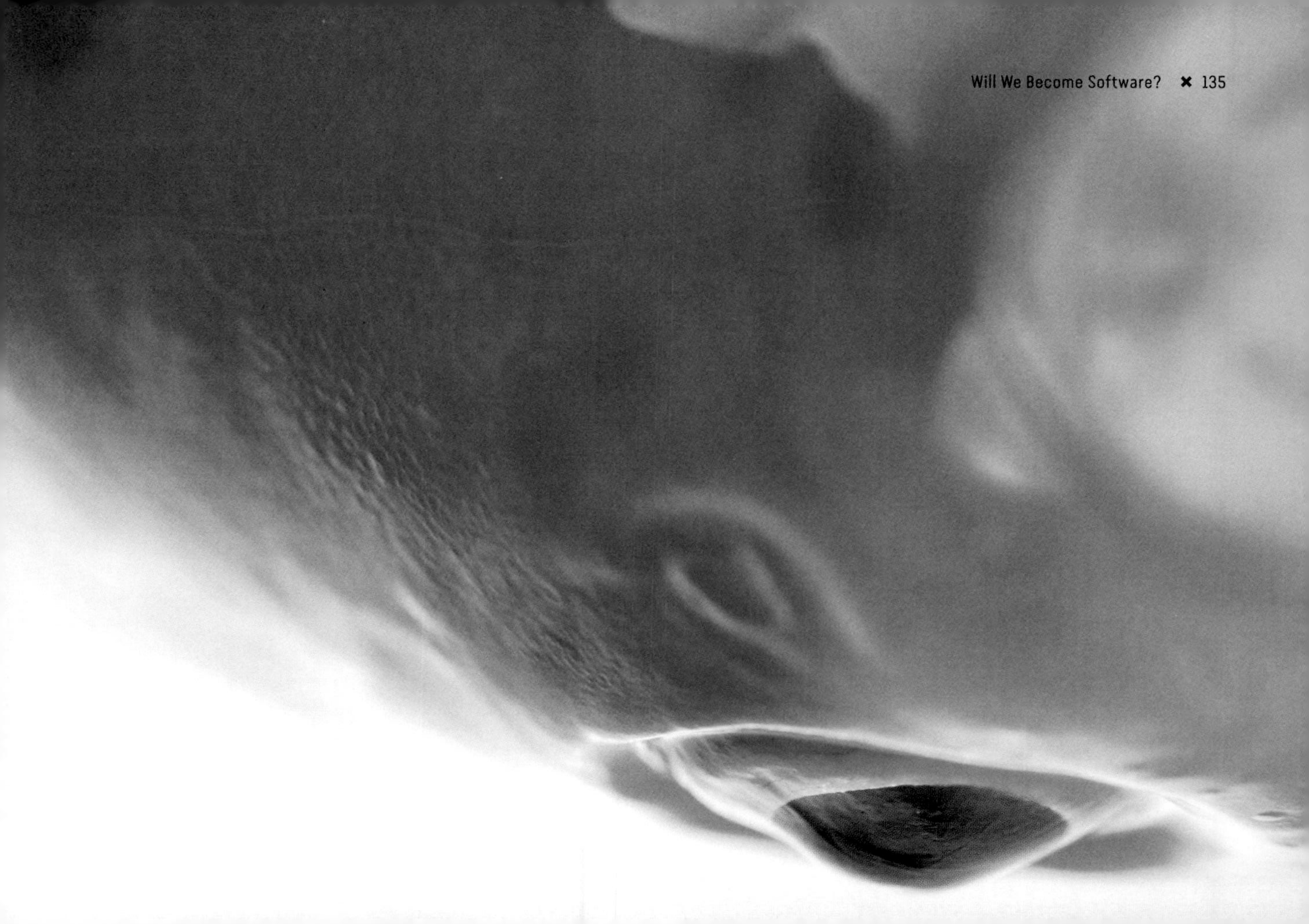

Will We Become Software? The Upload and Download

Today we bank all our fun into memories and curate our versions of ourselves on the Internet and social media, accessible to those we invite but also, largely, to the rest of the world. And you can't erase it. With your first post or like or share, you secure your immortality. In fact, the pursuit of digital immortality is becoming big business today, with tech pioneers creating algorithms that can scour the web to collect all your digital footprints and curate the living diary of you. And then there are apps that help you create that digital diary yourself, guiding you with questions that your loved ones may want to ask of you one hundred, five hundred, or even a thousand years from now.

One of the most interesting companies working on digital immortality is Terasem Movement Foundation, or TMF. TMF is the brainchild of Martine and Bina Rothblatt, and its focus is "mind uploading" through what they call "geoethical nanotechnology." They call the information captured "mindware" and aim to eventually reanimate it as conscious analogs.

Martine Rothblatt is a serial innovator, perhaps best known for cofounding satellite radio Sirius XM and the biotechnology company United Therapeutics. She and her wife, Bina, have set their sights on living beyond their biological boundaries. Working with David Hanson of Hanson Robotics around ten years ago, the couple created Bina48, a robot based on Bina's physical and psychological likeness, with a face that moves, eyes that see, ears that hear, and a digital mind that enables conversation. Bina48 is the first-ever robot to complete a college course and co-teach a university-level class (at the US Military Academy at West Point). Along with her "sister" Sophia, the first robot to receive citizenship (Saudi Arabia) and to be given a United Nations title (the first ever nonhuman Innovation Champion), they have reached humanoid celebrity status.

Sophia, modeled after actress Audrey Hepburn, was also created by Hanson Robotics. Both Bina48 and Sophia use AI, facial recognition, and voice recognition technology (speech to text), designed to get smarter over time. Their epigenetic clocks may be subject to wear and tear just like our meat bodies, but, unlike us, they have a greater chance of getting new parts and more advanced upgrades—indefinitely. One upgrade that is very near is an uncanny level of consciousness.

Japanese roboticist Hiroshi Ishiguro believes there are two levels of immortality: personal consciousness and social immortality. He is using technology to enable social immortality, which he describes as the ability to allow the dead to actively contribute to society from beyond the grave. Ishiguro has built an android version of himself and believes it could continue to teach robotics at Osaka University after he's dead.

MINDAR IS AN ADULT-SIZED android modeled after Kannon Bodhisattva, the Buddhist goddess of mercy. Designed with a gender-neutral body, it is programmed to deliver a twenty-five-minute sermon on the Heart Sutra, a Buddhist scripture, while moving its torso, arms, and head. Mindar was developed as a collaboration between the Kodaiji Temple and robotics professor Hiroshi Ishiguro of Osaka University. According to the priest Tensho Goto, "This robot will never die; it will just keep updating itself and evolving."

—*Japan Times*[1]

Who Owns Your Memories?

We won't all have the luxury of becoming a sentient android, but we can create our own "mindfile" that will represent us eternally. According to a blog post on Rothblatt's website, Lifenaut.com, most people in the world are compiling mindfiles, knowingly or not, through their unavoidable interface with digital communications systems (e.g., Facebook timeline, Google Glass, cloud auto-backups).[2] But what form would you like your mindfile to take after you are gone? One entrepreneur and researcher is building an application called "augmented eternity" that lets you create a digital persona that can interact with people on your behalf after you're dead. Hossein Rahnama, an associate professor at Ryerson University in Toronto and a visiting faculty member at the MIT Media Lab, is creating a text-based chatbot for an unnamed CEO that they both hope could serve as a virtual "consultant" when the actual CEO is gone.

One place you can dwell forever is on social media. According to one study, in as soon as fifty years from now (perhaps during the lifetimes of millennials and Gen-Zers), the dead are expected to outnumber the living on Facebook. The study's authors, Carl J. Öhman and David Watson, used a combination of data, including projected mortality and population rates from the UN, as well as Facebook's user growth over time.[3] Facebook now has hundreds of thousands of memorialized accounts, which are converted from a person's personal account once the company becomes aware of a user's passing from a friend or family member. On its website, Facebook states that it hopes to remain "a place where the memory and spirit of our loved ones can be celebrated and live on."[4]

Your views, your searches, your comments, and more are all part of your personal data. All of the digital immortality companies use machine learning that accumulates and analyzes every digital trace, but in theory, at least, they need your consent first. Öhman and Watson's paper, published

"Skype for the Dead"

Eternime collects your thoughts, stories, and memories, curates them, and creates an intelligent avatar that looks like you. This avatar will live forever and allow other people in the future to access your memories. Think of it like a "Skype for the dead," or a library where your children, friends, or even total strangers from a distant future will interact with the memory of you in a hundred years.

—Eterni.me

Luka and the Beginnings of Eternity in a Bot

Russian software developer Eugenia Kuyda created a chatbot representation of her best friend, Roman Mazurenko, who died in late 2015. Kuyda made the bot by plugging Mazurenko's personal messages to friends and family into a neural network built with Google's open-source machine-learning framework, TensorFlow. When the Luka bot answered questions, it sounded uncannily like her friend. Kuyda has since launched a company, Replika, to help people build a trusted friend in AI.

—Replica.ai

in the journal *Big Data & Society*,[5] opens with a quote from George Orwell's *1984*, which envisions a dystopian future of government surveillance and omnipresence. "We, the Party, control all records, and we control all memories," the quote reads. "Then we control the past, do we not?"[6]

Will we need a new set of ethics for our digital eternity? Carl Öhman and Luciano Floridi, from the Oxford Internet Institute, in a paper published in *Nature Human Behaviour*, say yes. They warn that we need to limit the ways in which companies can use (or exploit) our data. If digital remains are like "the informational corpse of the deceased," they write, they "may not be used solely as a means to an end, such as profit, but regarded instead as an entity holding an inherent value."[7]

Create Your Own Mindfile

Terasem Movement Foundation launched LifeNaut.com, which allows people around the world to store and share information such as text, files, images, photos, video, sounds, musical works, works of authorship, and other materials to create a computer-based avatar of themselves. Anyone with a connection to the Internet can create their own mindfile for free.

—LifeNaut.com

Mission Eternity

etoy.Corporation is a European digital arts group that launched Mission Eternity to collect, store and process information of human beings before and after they pass away. Timothy Leary (1920–96) left etoy.Corporation a large collection of documents and 32 grams of mortal remains that were integrated into the Mission Eternity system on May 26, 2007.

—MissionEternity.org

Black Mirror: "Be Right Back"

"Be Right Back," an episode of the British television show *Black Mirror*, is one of the most cited shows on digital immortality. It follows Martha after her lover, Ash, dies suddenly. Pushed into a special grieving service by a friend, Martha soon finds her deceased partner's social media activity deconstructed to duplicate an artificial, semi-constructed version of him. Over the course of the episode, Martha progresses from sending a few hesitant texts to a chatbot to purchasing a lifelike "Ashbot," a robot in her boyfriend's image. The avatar periodically elicits more and more of the dead boyfriend's data and upsells the grieving Martha on more expensive representations of him until it becomes so lifelike that she can't "kill" it. There's a controversial ending to this episode that we won't spill, but among fan sites, "Be Right Back" has been hailed as the best episode in the series.

—*Black Mirror* (season two, episode one)

Will We Consciously Live, and Love, Forever?

Martine Rothblatt's latest book, *Virtually Human: The Promise—and the Peril—of Digital Immortality*, explores the ways in which technology can help extend human life, and love, indefinitely. She imagines a universe populated by humans and their "mindclones." Mindclones are digital versions of people that live in parallel to the biological humans that created them. When the biological human expires, the mindclones live in perpetuity.[8]

Back in 2016, after movies like *Ex Machina* and *Her* had introduced the world to smart AI, Rothblatt was interviewed on the stage of South by Southwest, where she pragmatically explained that there will be continued advances in software, and eventually these advances will rise to the level of cognitive intelligence or artificial consciousness. At that point it will become very difficult to tell if the software is alive but without a body or if it is just fancy programming. She believes that "every other company is going to try to out-Siri Siri until we do have consciousness. And this is how I think it'll happen: it won't be like a giant announcement in the *New York Times*—'Today We Have Cyber Consciousness.' It'll be like water that just kind of rises and rises and rises, and before we know it, we're in an ocean of cyber consciousness."

But what really counts as "consciousness?" According to *Psychology Today*, the consensus in neuroscience is that consciousness is an emergent property of the brain and its metabolism.[9] So, when the brain dies, the mind—and the consciousness—of that person ceases to exist. Without a brain, there can be no consciousness. But the neuropsychiatrist Dr. Peter Fenwick disagrees, arguing that the brain does not create or produce consciousness; it simply filters it. Fenwick has been studying the human brain and the phenomenon of near-death experience (NDE) for fifty years. He believes that consciousness actually exists independently and outside of the brain as an inherent property of the universe itself, much like dark matter or gravity. And his views concur with other NDE studies that report that some people brought back from death after cardiac arrest claim to have experienced vivid memories and recollections even while their brain ceased to function.[10]

One case reported by Dr. Sam Parnia, a contributor to a study published in 2014 called AWARE (for "AWAreness during REsuscitation"), involved a man who essentially died of cardiac arrest and, when resuscitated, accurately described the people, sounds, and events of his "resurrection."[11] Evidence from AWARE and other studies raises the possibility that the mind, or consciousness, may not originate in the brain. It may be a separate entity. Some ancient cultures believed that consciousness is not part of the body but part of nature and the universe, connected to frequency and the electromagnetic waves that carry sound. We just don't have the modern tools to measure this—yet. Parnia and his team at NYU Langone Health are conducting further NDE studies on people in cardiac arrest to see what happens to the mind after our body dies—that is, if our consciousness is actually immortal.

Are You Ready for Your Upgrade?

With the potential to live longer, the first frontier for improvement will be the brain. We are seeing a swarm of investors expanding their interests in digital health technologies toward the emerging technology that blends computers and biology. The new field of neurotechnology, estimated to reach $13.3 billion in 2022 (Research and Markets), is led not just by the established medical labs but start-ups with some serious backing. Neurotechnology includes brain-machine interfaces (or BCIs), and neuroprosthetics, implants to enhance memory and cognition, neurostimulation, and neuromonitoring. The first applications of neurotechnology include treatments for spinal cord injury, deafness, blindness, stroke, epilepsy, chronic pain, and psychiatric disorders.

> FACEBOOK IS MOVING BEYOND LIKES and shares to capturing your thoughts. Its newly formed brain-computer interface (BCI) research program at Facebook Reality Labs aims to develop a noninvasive, silent speech interface that will let people type just by imagining the words they want to say—a technology that could one day be a powerful input for all-day-wearable augmented reality (AR) glasses.
>
> —Tech@Facebook

If you've ever read a William Gibson or Neal Stephenson novel, you've had a peek at what BCI can do. Some big players are banking on BCI interfaces that could someday lead us into a mind-controlled world. Elon Musk, the billionaire polymath who gave us Tesla Motors, SpaceX, and the start of PayPal, has a new, somewhat secretive company called Neuralink, which is developing "ultra-high bandwidth brain-machine interfaces to connect humans and computers."[12] Facebook Reality Labs is making progress on its brain-machine interfaces, which would allow users to type using their thoughts.

The Brain-to-Brain Hookup

A report by the Royal Society estimates that by 2040 neural interfaces will be an "established option" for effectively treating diseases like Alzheimer's.[13] But future benefits also include supersensory capabilities: brain implants that allow people to virtually taste, smell, and see. The report also details implants that could boost memory, improve vision, and even allow thoughts to be transmitted from one person to another. Like Neal Stephenson and J. Frederick

ACCORDING TO ITS WEBSITE, KERNEL is a company "building a non-invasive mind/body/machine interface (MBMI) to improve, evolve and extend human cognition." It was founded by Bryan Johnson, founder of the online payments company Braintree and the OS Fund.

—Kernel.co

BRAINCO, INCUBATED IN THE Harvard Innovation Lab, develops brain-machine interface (BMI) technology products, including sensors, hardware, software, and AI. Its Focus series offers wearable headbands for education (FocusEdu), fitness (FocusFit), and mind-controlled games.

—BrainCo.tech

George's character in *Interface*, a presidential candidate with the competitive advantage of having the mood of the electorate channeled directly into his brain via a biochip, the Royal Society report warns that eventually "people could become telepathic to some degree, able to converse not only without speaking but without words, through access to each other's thoughts at a conceptual level. This could enable unprecedented collaboration with colleagues and deeper conversations with friends."[14]

Mary Lou Jepsen, a former MIT professor and engineering executive at Facebook, Oculus, Intel, and Google, claims it should take fewer than eight years to make telepathy possible. Her company Openwater is developing a device that puts the capabilities of a huge MRI machine into a lightweight, wearable form, like a thinking cap. Her hope is that communicating by thought alone wouldn't just be faster but would allow humans to compete effectively with the artificial intelligence that will be infiltrating our lives.[15]

The Defense Advanced Research Projects Agency (DARPA) has been one of the leading institutions in the field of brain-computer interfaces since the 1970s. DARPA coined the term "operational neuroscience" to describe the combination of neuroscience, cognitive psychology, and physiology that investigates how to maximize human potential. A study DARPA funded reaffirmed that transcranial direct current stimulation, a painless brain stimulation treatment that uses direct electrical currents to stimulate specific parts of the brain, can lead to "widespread changes in brain activity" and suggests that it can be a low-cost, noninvasive way to alter brain functionality and optimize cognitive abilities. The US Department of Defense is creating an implantable brain device that restores memory-generation capacity for people with traumatic brain injuries. Published study results report that their patients showed memory retention improvement of as much as 37 percent.[16]

What's your brain age? We are getting closer to the point where our brain waves and body rhythms will be traceable and malleable. One of the early indicators of aging in the brain is a slowing of our dominant alpha brain wave frequency. However, neuroscience suggests that not all brain functions deteriorate with aging; some continue to mature as we age. For example, it is thought that we experience peak ability to concentrate at forty-three, peak ability to recognize emotions at forty-eight, and peak vocabulary ability at sixty-seven.[17]

There are dozens of apps to help you safely stimulate your brain. Known as "brain entrainment," these therapies use either light, electromagnetic waves, or pulsing sound, also known as binaural beats. Binaural beat therapy is an emerging form of sound wave therapy in which the right and left ears listen to two slightly different-frequency tones yet perceive the tone as one. It is likened to the effect of meditation.

More sophisticated therapies aim to measure and treat the aging in your brain. Haruo Mizutani, chief scientist at PGV, a Tokyo-based tech company, demonstrated a high-precision, miniaturized brain wave sensor for measuring "brain age" at Teijin's The Next 100 Think Human Exhibition. Biocybernaut Institute offers alpha brain wave training

AT THE AGING PROJECT BOOTH, PGV measured visitors' "brain age," which they displayed in real time on a monitor. They also displayed charts showing correlations between age and peaks in brain functioning.

—Teijin Next 100 Think Human Project

Twenty-One to Forty Years of Zen in Seven Days

Biocybernaut brain wave training uses state-of-the-art eight-channel EEG neurofeedback technology systems combined with coaching and instructions from certified Biocybernaut trainers. The unique seven-day Biocybernaut brain training programs can make twenty-one to forty years of progress in just one week.

—Biocybernaut.com

NEUROSCAPE IS VIDEO GAME RESEARCH to improve mind quality that uses a cutting-edge approach to improving brain function, building a bridge between neuroscience and consumer friendly technologies. Video games are being developed to support treatment of brain disorders such as ADHD, autism, depression, multiple sclerosis, Parkinson's disease, and Alzheimer's disease.

—Neuroscape.ucsf.edu

to develop latent abilities in older people as well as to reverse the effects of aging in their brains. The company claims that in seven days you can make the same changes in your brain waves as you would spending twenty to forty years in a serious meditative practice.[18]

Man is not going to wait passively for millions of years before evolution offers him a better brain.

—Corneliu E. Giurgea, Romanian scientist, who coined the term "nootropic" in 1972

Neuro-enhancements aren't all devices and implants. Remember the 2011 movie *Limitless*, about a man who takes a special pill and becomes smarter and more capable than anyone else on earth? Big money is pouring into consciousness research as well as psychedelics, a class of drugs that produce profound changes of consciousness over the course of several hours. The Johns Hopkins School of Medicine recently launched the Center for Psychedelic and Consciousness Research, the first and largest of its kind, focusing on how psychedelics affect brain function, learning, and memory. Then there is the biohacker's arsenal of nootropics—also known as "smart drugs" or "cognitive enhancers"—that offer some kitchen-meets-lab remedies for supercharging your brain power. From chewables and dietary supplements to "brain octane" oils like MCT made from a type of fat called medium-chain triglycerides that turn into energy you can use when added to your daily caffeine fix, nootropics are enjoying a comeback. The theory is to keep your brain stimulated for peak performance and to combat aging.

So are we close to an eternal mind? Some experts believe we are still years away, even though the digital immortalists are working to bank your brain in software. The problem is that first you have to be dead and your brain has to be preserved in microscopic detail using a high-tech embalming process so that down the line scientists can scan it and turn it into a computer simulation.

In an interview with Neo.Life, geneticist George Church poses an analogy between brains and computers that is worth considering as we venture into brain augmentation. He alludes to the likely desire to have a more flexible brain size so you can add components or more working memory—much as you would add RAM to a computer.[19] But as we aim to become smarter, he says, we have to ask, "Are we maintaining the neurodiversity we need?"

Would a society of exponentially smart people, both living and digitally represented, be a good thing? "Neurodiversity," a word first popularized in the late 1990s by Australian sociologist Judy Singer and American journalist Harvey Blume, refers to the variation of how each of us thinks and therefore acts in the course of sociability, learning, attention, mood, and other mental functions. It also determines how we sense the world around us.

Scientists today believe we have more than twenty-one senses with which to experience the world around us. Some new technologies can help us tap into these innate senses; one is magnetoreception, which is the ability of fish, birds and other migratory animals and insects to detect the earth's electromagnetic fields. A team from Lund University in Sweden recently found that humans can also perceive magnetic fields, though unconsciously, since changes in magnetic fields affect our alpha brain waves.[20] (Stay tuned for more on the amazing power and potential of frequency in our next book, *Tuning Into Frequency*.)

Our memory is feeding our future. Memory is a storage bank of virtual possessions. Today a synesthetic culture is taking form. Most of the synesthetes I know are more plugged into their environment; they are just more plugged into their surroundings, more than other people. It can even be said that the synesthete is even more unique than the non-synesthete. The synesthete's mind has a remarkable plasticity to hear color, see sounds, and taste tactile sensations. In a sense, synesthetes are the people of the future. Because, with the futuristic, telematic extension of the human senses, everything will become more and more synesthetic. The senses will become telesenses. Telesensing is the integration of the visual, the kinetic, the telematic, and the synesthetic. Coming is a telematic culture. In the near future we are going to see proof that the ultimate software that we can use is what they call "wetware" [hardware software, which] will have a telematic quality, which is a mind-to-mind connection. Telepathy is the ultimate software and that will become real, in fact. In the near future this kind of brain-to-brain and mind-to-mind contact will be possible in a telepathic way.

–Hugo Heyrman, Belgian-Flemish painter and synesthetic artist;
interview with Sputnik Futures

Supersapiens

In the 2017 film *Supersapiens, the Rise of the Mind*, writer-director Markus Mooslechner raises a core question: As artificial intelligence rapidly blurs the boundaries between man and machine, are we witnessing the rise of a new human species? "Humanity is facing a turning point—the next evolution of the human mind," notes Mooslechner. "Will this evolution be a hybrid of man and machine, where artificial intelligence forces the emergence of a new human species? Or will a wave of new technologists, who frame themselves as 'consciousness-hackers,' become the future torch-bearers, using technology not to replace the human mind, but rather awaken within it powers we have always possessed—enlightenment at the push of a button?"

—Kurzweil Network

So, if in the future we can all choose to have higher cognitive operating systems, put on our telepathic reading caps, instantly copy thoughts to our cognitive clouds, and download new emotions at will, could this be the dawn of our evolution from a biology of cells to a biology of cells, binary bits, and qubits?*

*A binary bit is a basic unit of information in information theory, computing, and digital communications (https://en.wikipedia.org/wiki/Bit). A qubit, or quantum bit, is the basic unit of quantum information—the quantum version of the classical binary bit (https://en.wikipedia.org/wiki/Qubit).

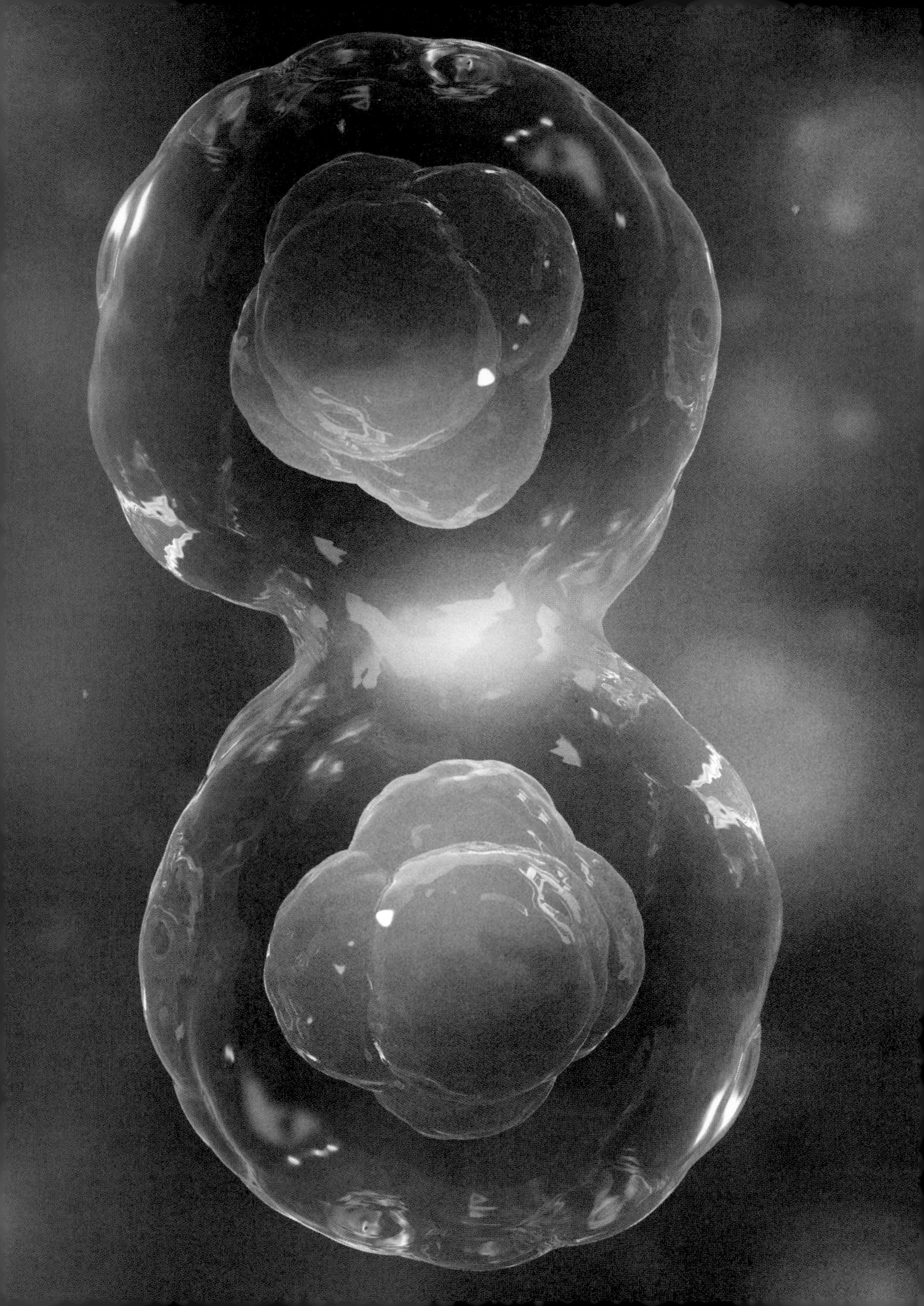

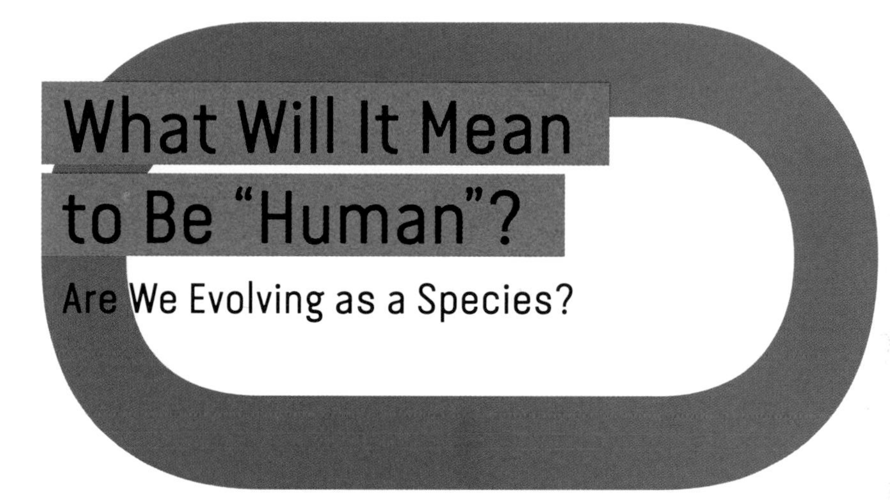

What Will It Mean to Be "Human"?

Are We Evolving as a Species?

Creativity is a human gift. And we can use this creativity to alter our own genetics and those of other critters, and soon design entirely new ones.

It is possible that evolution will be driven by human choices, and as a result we will see the creation of an entirely new species of human being. Will this new human be, well, a more "human" human? Or will we (black) mirror the dystopian Hollywood future? Today we have the longevity players: the cryonicists (freeze and reanimate), the extropians (freedom from all limitations), the transhumanists (body enhancements), and the Singularitarian mind uploaders (uploading the mind into a computer). Perhaps all these players will be forerunners to a new group called the "humane" humans. To quote People Unlimited, a community of people who question the limitations of aging and death, "People who are preparing to live rather than to die are a lot more fun to be around." And according to social network expert Nicholas Christakis, emotions are contagious. So what's the best way to be human? The best answer, supplied by the Greater Good Science Center's founder Dacher Keltner, is for this new humane human to practice compassion.

Meet the Immortality Players

THE CRYONICISTS: The goal of cryonics is the low-temperature freezing and storage of the human body with the hope that it can be resurrected in the future. It is a way to preserve you for decades until a future medical technology can restore you to full health.

THE EXTROPIANS: Extropians hold the belief that advances in science and technology will someday let people live indefinitely. "The Principles of Extropy,"[1] a set of ideas originated by the philosopher Max More, places strong emphasis on an optimistic view of the future in expectation of considerable advances in computational power, life extension, nanotechnology, and more. In general, extropians foresee the eventual realization of indefinite life spans, thanks to future advances in biomedical technology or mind uploading.

TRANSHUMANISTS: Transhumanism, abbreviated as H+ or h+, is an international philosophical movement that advocates for the transformation of the human condition by developing and making widely available sophisticated technologies to greatly enhance human intellect and physiology. Transhumanists are the original biohackers, upgrading their bodies and minds through technology such as implants to enable them to exceed normal physical and mental limitations.

SINGULARITARIAN MIND UPLOADERS: Singularitarians believe in whole brain emulation (WBE), mind uploading, or brain uploading (sometimes called "mind copying" or "mind transfer"), a process of scanning long-term memory and the thoughts that make the "self" and copying them onto a computer.

Ray Kurzweil, author of *The Singularity Is Near: When Humans Transcend Biology*, talks about the coming Singularity (estimated to be around 2045) that will mark the moment when artificial intelligence becomes smarter than humans. Singularitarians believe it will be prudent to upload your mind—and your soul—to achieve immortality.[2]

People Unlimited

An educational, lifestyle and social organization, People Unlimited provides information for those interested in living unlimited life spans. Its Ageless Education series brings in many of the leading figures in radical life extension to share their views on the most cutting-edge strategies for living long enough to live forever.

—PeopleUnlimited.com

What Makes Us Human? Some Say Our Superior Intellect; Others Argue It's Our Emotions

Emotions are one of the biological processes that are still not easily quantified. In recent years we have learned that many emotions are not created purely by our mental state or our nervous system. Emotions are induced by chemical changes and can be influenced by thoughts, feelings, and behavioral responses. Scientists are now finding that subjective emotions, and the degrees to which we feel them, can influence our aging. For example, anger can shorten your life; prolonged periods of anger can especially affect your cardiovascular health, but new findings show that it can also create all kinds of health problems via chronic inflammation. By measuring our blood's C-reactive protein levels that are a marker for inflammation, we can see the damage to our immune response, which is crucial to holding off those age-related diseases that keep coming up: osteoarthritis, diabetes, some cancers, heart disease, and even dementia. On the other side of the spectrum, positive feelings like forgiveness and compassion may boost our immune system and help improve cardiovascular and endocrine functioning.

Quantifying our emotional influence on our immune system and its impact on aging is still a new field, known as psychoneuroimmunology (PNI).

Steve Cole, a professor of medicine, psychiatry, and bio-

Healthful Advice: Practice Radical Optimism

There is growing evidence that optimistic men and women may live longer than those who are pessimistic. (We at Alice in Futureland are self-described eternal optimists!) There have been early studies that tested the effects of adopting an optimistic view, and the results have shown that more optimistic individuals are less likely to suffer from chronic diseases and die prematurely.

behavioral sciences at the UCLA School of Medicine, researches the biological pathways by which social environments influence gene expression by viral, cancer, and immune cell genomes. Earlier in his career he published studies suggesting that negative mental states, such as stress and loneliness, affect our immune responses by triggering the mechanisms that lead to gene expression, shaping our ability to fight disease. PNI studies have mostly looked at levels of individual immune cell types; molecular messengers such as the stress hormone cortisol and the immune messenger proteins called cytokines; or the expression of individual genes.[3] In the study "Stress, Age, and Immune Function: Toward a Lifespan Approach" by Jennifer E. Graham, Lisa M. Christian, and Janice K. Kiecolt-Glaser, published in the *Journal of Behavioral Medicine* in 2006, the authors explain that psychological stress can both mimic and exacerbate the effects of aging, with older adults often showing greater immunological impairment to stress than younger adults. In addition, stressful experiences very early in life can alter the responsiveness of the nervous and immune system.[4]

RESEARCHERS AT BOSTON UNIVERSITY SCHOOL of Medicine recently shared amazing results suggesting that optimism is specifically related to "11 percent to 15 percent longer life span, on average, and to greater odds of achieving 'exceptional longevity'—that is, living to the age of eighty-five or beyond." The relation of optimism to life span persisted independent of socioeconomic status, health conditions, depression, social integration, and health behaviors (e.g., smoking, diet, and alcohol use). Their findings suggest optimism may be an important "psychosocial resource for extending life span in older adults."[5]

THE BOSTON UNIVERSITY SCHOOL of Medicine research furthers the work of a 2002 study published in the *Journal of Personality and Social Psychology* that suggested that "having an optimistic mind-set about aging could add a full seven and a half years to your life." PLOS ONE reported in 2018 that other studies have linked an individual's subjective age (how old you feel relative to your chronological age) and their aging mind-set (what qualities you associate with old age) to the risk of developing dementia. According to the results of the BU study, an optimistic aging mind-set can in fact protect against dementia, even among those who carry APOE4, a variant of the APOE gene, the strongest known risk factor for late-onset Alzheimer's.[6]

Survival of the Kindest

Dacher Keltner has spent much of his career studying human emotions, particularly the "social" emotions, like gratitude, amusement, awe, and compassion. A professor of psychology at the University of California, Berkeley, and founder of the Greater Good Science Center, his research focuses on the biological and evolutionary origins of compassion, awe, love, beauty, emotional expression, power, social class, and inequality. In an interview with Sputnik Futures, Keltner explains that when you're in these emotions of awe, gratitude, compassion, beauty, and wonder, it is almost impossible to be physically stressed. This is especially true for compassion and kindness.

His work shows how positive emotions—which contribute to the feeling of optimism—can affect our immune system. He explains in our interview how positive emotions are linked to the vagus nerve, the main nerve of the parasympathetic division of the autonomic nervous system, which influences immune responses and inflammation, and how practicing kindness promotes positive health and an altruistic ripple effect:

We have emotions in our bodies—compassion and gratitude and awe and beauty and wonder—that actually are connected up to dopamine circuits of the brain, which are part of reward and help build addictive behavior. We know those emotions are intimately connected to vagal response, vagus nerve, immune system processing, and positive health profiles. We know when you're in these emotions, like gratitude or compassion, it is almost impossible, physically, to be stressed.

We are really learning that we aren't separate individuals; that we aren't separated by space and bounded by skin; that we're really actually very connected to each other. And that is just a very emergent idea and a defining characteristic of human evolution, that we are this profoundly mimetic species where we imitate each other, and we take on each other's attributes and emotions. And new science is revealing that is especially true for compassion and kindness. There is work showing that if I'm generous to you or kind to you, that act of kindness you take with you, away from that context, promotes kind behavior toward others, and it ripples out and spreads four or five more times.

—Dacher Keltner, social psychologist, founder of Greater Good Science Center, author of *Born to Be Good: The Science of a Meaningful Life*, host of the podcast *The Science of Happiness*; interview by Sputnik Futures

Characteristics of Social Emotions

Awe is that wonderful moment when you pause, struck to your core by the beauty or meaning of something. Awe shifts a person's thinking toward the collective and engenders a sense of humanity.

Gratitude is when you express thanks for a gift that's given to you, or a trade or cooperation. It can be through touch, eye contact, words, or tone of voice. Gratitude is the emotion of social reward and an engine of cooperation.

Compassion is the foundation of our moral sense and the essence of cooperative communities. When people see others respond with compassion, they feel inspired and uplifted.

—Greater Good Science Center

We all exhibit the same social emotions. Interestingly, no matter where you live or your cultural background, we all share the same facial expression when in a state of positive emotion. Through the facial recognition studies that began with Paul Ekman, the American psychologist whose work inspired the television show *Lie to Me*, Keltner and Ekman have studied how our biological predisposition is to be compassionate. Therefore, if positive social emotions can support a healthier immune response, perhaps we should add a daily dose of wonder and gratitude to our list of antiaging supplements.

As populations grow, social emotions will be important for the new social contracts we will be making not only among ourselves but between us and the artificial intelligence we will cohabitate with. Our social emotions act like a biological meme, connecting us and going viral through our networks of friends and our friends' friends. Nicholas A. Christakis has been at the forefront of proving that our emotions are contagious through our networks and that we are hardwired to connect through emotions and behavioral influences. The seed of this theory began with

the Framingham Heart Study, in which Christakis and James Fowler did a longitudinal social network analysis over twenty years to evaluate whether happiness can spread from person to person and whether niches of happiness form within social networks.[7] They used as their sample more than four thousand individuals within the Framingham Heart Study (FHS) social network, following them from 1983 to 2003. They concluded that people's happiness depends on the happiness of others to whom they are connected. This suggests we should think of happiness, like health, as a collective phenomenon.

Interestingly, one of the earlier studies on aging and exceptional longevity also used data from the FHS, which began in 1948, involving more than five thousand individuals between the ages of twenty-eight and sixty-two in the town of Framingham, Massachusetts.[8] Drawing on this data, a team from Duke University's Center for Demographic Studies used longitudinal data from the physiological index and life spans collected from the FHS participants, analyzing indexes that can affect longevity and age-related

NICHOLAS A. CHRISTAKIS, MD, PHD, MPH, is a sociologist and physician who conducts research in the areas of social networks and biosocial science. He directs the Human Nature Lab. His current research is mainly focused on two topics: (1) the social, mathematical, and biological rules governing how social networks form ("connection"), and (2) the social and biological implications of how they operate to influence thoughts, feelings, and behaviors ("contagion").

diseases such as blood pressure, body mass index, serum cholesterol, and blood glucose (some of the biomarkers we discussed in chapter 02). They found that the behavior of these physiological indexes at the critical ages of forty to sixty when at risk for onset of age-related diseases affected the late-age survival of the individuals. The 2006 study raised the importance of such information for better understanding of the aging process and age-associated changes in health and well-being.[9]

The FHS study has also been a rich data source for studying a regional cohort (not a massive blind study), contributing to important work in epigenetic and genetic prediction of heart disease and region-based DNA methylation analyses for novel biomarkers and mechanisms of cardiovascular disease. Again, where you live impacts your chances of reaching the ripe old age of one hundred merely because you and your neighbors are *connected* by proximity and lifestyle. Which begs the question: If happiness and health are contagious, could *longevity* be too?

Speaking of neighbors, it's time we humans face up to the fact that AI will be our friends, caretakers, coworkers, family members, and neighbors. In his latest book *Blueprint: The Evolutionary Origins of a Good Society*, Nicholas Christakis uses his network thinking research to prove that we are genetically wired to create good societies—that our genes affect not only our individual bodies and behaviors but also the shape of our societies. In an article for the *Atlantic*, Christakis questions the new social contracts we will be making with AI and whether that new feedback loop will rewire humans' capacity for altruism, love, and friendship—the very inherited capacities that help us live communally. He believes that "unfortunately, humans do not have the time to evolve comparable innate capacities to live with robots" and we will need to take steps to make sure we are not destructive "as AI insinuates itself more fully into our lives." His view is that "for better and for worse, robots will alter humans' capacity for altruism, love, and friendship."[10]

Science is delivering on its promise to upgrade our bodies and our minds, and with this comes a range of ethical and social questions. We won't just be making new social contracts with AI; we will be stretching the notion of what it means to be "human." Some experts believe that we haven't yet fully realized our true human potential, and we are about to embark on an extraordinary evolutionary shift toward becoming a more "human" human. The field of human augmentation—sometimes referred to as human performance enhancement, or "Human 2.0"—focuses on creating cognitive and physical improvements as an integral part of the human body. Often the goal is to achieve superhuman potential and to bolster extreme longevity.

Zoltan Istvan, an American transhumanist, entrepreneur, and futurist, explains in *Psychology Today* that whether you use the term "singularity," "posthuman," "techno-optimism," "cyborgism," "immortalist," "biohacker," "life extension," or "transhumanism," they are all terms being used "to describe a future in

which mind uploading, indefinite lifespans, artificial intelligence, and bionic augmentation may (and I think will) help us to become far more than just human."[11]

Perhaps more important, in our quest for better performance and infinite life spans, we are faced with new ways of thinking about the "self." In an interview with Sputnik Futures, Erik Davis, an American journalist, talks about the idea that we are now shapers of our own being:

There's an interesting quote; this is a supreme being describing this thing that he's done to man: "We have made you [humans] a creature neither of heaven nor of earth, neither mortal nor immortal, in order that you may, as the free and proud shaper of your own being, fashion yourself in the form you may prefer." Sounds pretty similar. This is from the fifteenth century, Pico della Mirandola and his humanist creed on the oration of man. What a miracle is man. It is very interesting because it shows us that a lot of these issues we face about how we're transforming ourselves, what we're becoming, how we're forming, what values are guiding this, are very, very old. So there's nothing very new about the idea that we're shaping ourselves, and now we can do things that are going to shape ourselves in new ways, and in some sense this is a free process in that it's, in fact, the freedom of the process that calls us further into it.

What Will Be the Consequences of Pushing Our Human Limits to Cheat Death?

We are code. The code of life has a structure and a sequence, which geneticist George Church and others found in the Human Genome Project. Our genetic and epigenetic sequencing are the first clues toward how to read this code of life. We now have CRISPR and other gene editing tools and are at the brink of being able to write, and rewrite, the code of life ourselves.

With this code and our editing tools, we will become "programmable." Programming life extension, as opposed to allowing for the biological and natural selection process of decay and death, is the driving passion of human augmentation proponents. In their view, the fusion of biological and machine parts is natural, since our bodies are tools just like cars and computers. They can be fixed, renewed, and upgraded to extend life—and improve operations. Zoltan Istvan explained in an interview with Sputnik Futures that "there are always chemical reactions happening in our body, billions at once, that are creating who we are. And the real question is, how can we get hard machine ware to do that exact same thing, especially on the molecular level? We're going to get there with nanotechnology. It's just a little bit longer in taking and I think everyone wants to get there."

If this sounds far out, consider that for the past few decades we have already been riding the tide of human enhancement. We already have alertness pills, athletic and energy boosters, sexual performance enhancers, immunity boosters, prosthetics that replace limbs, organ transplants, bone implants, retinal and cochlear implants, and plenty more. Then we have the wading pool of nootropics, supplements, psychedelics, and fragrance molecules that can facilitate trust, encourage pair bonding, improve sleep and memory, and alleviate pain.

There is a disconnect between the dystopian visions popularized in the media and the pioneers on the front lines of human augmentation. Istvan and others believe this debate will change, and with the introduction of more lifesaving technologies we will see greater social acceptance. If you can get a new hip at the age of seventy-five, why stop there?

I AM HUMAN IS A FILM that asks: Who are we? What might we become? And, perhaps most important, are we ready? *I Am Human* highlights several of the world's first "cyborgs"—a tetraplegic, a blind man, an artist with Parkinson's—and the scientists and entrepreneurs on a quest to unlock the secrets of the brain.

—IAmHuman.com

FOR TEIJIN'S THE NEXT 100 Think Human Exhibition, the humanness project team considered the relationship between humanity and the technologically transformed human body in an effort to present insights that would illuminate the future.

—100.teijin.co.jp

We will eradicate disabilities in our quest to live longer and healthier. We might even be able to put together superhuman bodies.

When we realize our augmented potential, the age associated with our physical bodies won't carry much meaning. Professor Masahiko Inami of the Research Center for Advanced Science and Technology at the University of Tokyo considers transitions in our sense of "humanity" from the perspective of our bodies. He states, "There have been reports of psychological studies showing that changing our bodies also changes our minds, such as becoming kinder after being recast as a superhero . . ."[12] Today we are already seeing the use of augmenting devices to help the paralyzed walk or to help those with limited strength or range of motion in their hands with a robotic glove.

DARPA, the Defense Advanced Research Projects Agency of the United States Department of Defense, is one of the federally funded organizations leading the charge to conquer human augmentation. We can thank DARPA for providing significant technologies that influenced the development of computer networking and the basis for the modern Internet, as well as graphical user interfaces in information technology. (Side note here: The agency was created in February 1958 by President Dwight D. Eisenhower in response to the Soviet launching of Sputnik 1 in 1957. No wonder we at Sputnik Futures–the human team of Alice in Futureland–are big fans!)

The latest line augmentation research at DARPA, the Next-Generation Nonsurgical Neurotechnology (N^3) program, is looking at creating a sort of "short-term superhuman."[13] Its focus is on one key part of augmenting soldiers: making sure the effects can be reversed. Creating a seamless interface between human and machine invokes images of brain implants and cutting-edge prosthetics. The webpage of DARPA's program notes that "the most effective, state-of-the-art neural interfaces require surgery to implant electrodes into the brain." But we may not want these neural interfaces to be permanent, especially if they are programmed to mentally survive exceptional physical endurance. The N^3 program is part of the current trend in US military research: figuring out temporary, noninvasive ways to enhance soldiers.

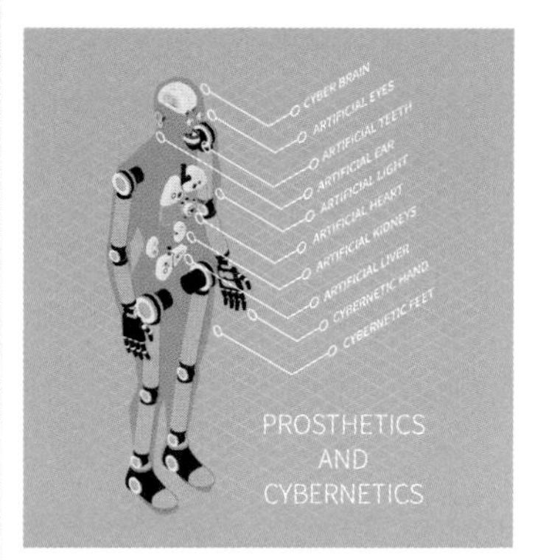

Wearable Superpowers

Roam Robotics is making powered human augmentation devices to enhance human mobility in activities such as walking, hiking, skiing, biking, running, and jumping. Its goal is making a world possible where extraordinary physical experiences of strength, endurance, and speed are accessible to the average person.

—RoamRobotics.com

Cryonics: Get Ready for the Long Float

If you do choose to upgrade and enjoy an additional twenty to thirty years of healthy living, you may have another option for planning your immortality: cryonics. If you saw the movie *Interstellar* or any sci-fi space travel movie, the astronauts use some form of cryonics in their journey.

Cryonics is an effort to save lives by using temperatures so cold that a person beyond help by today's medicine can be preserved for decades or centuries until a future medical technology can restore that person to full health.

The idea that a person could be frozen and then brought back to life when the technology had evolved far enough originated with the book *The Prospect of Immortality*, written by physics teacher Robert C. W. Ettinger in 1962. The word "cryonics" is derived from the Greek term for "cold."

One company banking on the science that will enable us to preserve ourselves for a future when we can live (potentially) forever is Alcor Life Extension Foundation, which is currently taking applications for membership, so that if you become terminally ill, you can avoid death and have a tank ready for you at Alcor. They even supply you with an emergency ID tag for a bracelet or necklace to inform them in the event of emergency care.

The first person to be cryogenically frozen was a seventy-three-year-old psychologist, Dr. James Bedford, who was suspended in 1967. His body is reportedly still in good condition at Alcor Life Extension Foundation.

While the reality of what happens to the human body after a suspended long sleep is still unknown, scientists and engineers are collaborating with NASA and other space agencies to develop suspended animation options for missions to Mars and beyond. Their approach is not a deep freeze but keeping the astronauts in a long, induced sleep called torpor that acts like biological suspension for weeks or months. The hope is that this process will allow the spaceships to be more compact, with less equipment, and therefore lighter to propel through space. Experts believe that if this experiment is successful, it will also help to maintain the physical and mental health of the space travelers. And, of course, torpor could help us here on Earth too.

The Cryonics Premise

According to the Alcor Life Extension Foundation, cryonics is justified by three facts that are not well-known:

1. Life can be stopped and restarted if its basic structure is preserved. Human embryos are routinely preserved for years at temperatures that completely stop the chemistry of life.

2. Vitrification (not freezing) can preserve biological structure very well. Adding high concentrations of chemicals called cryoprotectants to cells permits tissue to be cooled to very low temperatures with little or no ice formation.

3. Methods for repairing structure at the molecular level can now be foreseen. The emerging science of nanotechnology will eventually lead to devices capable of extensive tissue repair and regeneration, including the repair of individual cells one molecule at a time.

—Alcor.org

The Noah's Ark of Immortality

When built, Timeship will be the world's most secure and technologically advanced facility for the long-term storage of cryopreserved biological materials and a center for research into life extension. Its six-acre structure has been designed to provide safety and security for up to one hundred years against multiple threats, both natural and human-made, including terrorist attacks, changes to sea levels, and the interruption of energy supplies.

—Stephen Valentine, *Timeship: The Architecture of Immortality*

h+

Halle's accessories:

Data display contact lens

Configurable permanent makeup

Personal cell, PDA facial stud

Solar, climate controlled hoodie

Probing de Grey Matters

AUBREY DE GREY

The Reluctant Transhumanist

CHARLIE STROSS

Don't Leave Your Memory at Home

BUILDING BETTER BRAINS

Billionaires Funding the Future

SCIENCE FICTION GETS FUNDING

Editon #1
Fall 2008

Emerging and speculative technologies will change the way we live. Transhumanism is not exactly a new idea. The word "transhumanism" was first used by Aldous Huxley's brother, Julian Huxley, a distinguished biologist who was also the first director general of UNESCO and a founder of the World Wildlife Fund.

In *Religion Without Revelation* (1927), Julian Huxley wrote: "The human species can, if it wishes, transcend itself—not just sporadically, an individual here in one way, an individual there in another way, but in its entirety, as humanity. We need a name for this new belief. Perhaps transhumanism will serve: man, remaining man, but transcending himself, by realizing new possibilities of and for his human nature."[14]

Nick Bostrom, professor of applied ethics and director of Oxford University's Future of Humanity Institute, is known for his work on existential risk, the anthropic principle, human enhancement ethics, superintelligence risks, and the reversal test. In 2005 he published one of the most widely respected papers on transhumanism, called "A History of Transhumanist Thought."[15] He states that "the human desire to acquire new capacities is as ancient as our species itself," and that some individuals always "search for a way around every obstacle and limitation to human life and happiness." Transhumanists believe in the potential for genuine improvements in human well-being and human flourishing, explains Bostrom, and that these are attainable only through technological transformation.

According to Dr. Natasha Vita-More, one of the foremost thought leaders of transhumanism, the worldview aims to prepare us for the challenges of the exponential leaps in technology that are coming in the future. Rather than playing God, humanity needs to be informed about advances in biotechnology, artificial intelligence, and nanotechnology. Transhumanists think that the ethical use of technology and evidence-based science will help each person to determine their own evolution. Vita-More uses the phrases "life extension" and "life expansion" rather than "immortality," because she thinks "immortality" can be confused with "fantasy," and transhumanist thinking is based on critical analysis of possibilities for extreme longevity. In her view, life is about experiencing challenges that motivate us to reach new levels. In an interview on Medium, Vita-More suggests that if you could live indefinitely, there would be ample possibilities to explore new ideas. The myths of Icarus and Prometheus would be folklore and Van Gogh's struggles would be a condition of the past.[16]

One Russian billionaire is so convinced immortality is inevitable that he's created and funded his own nonprofit to make it a reality. According to 2045.com, Dmitry Itskov's 2045 Initiative team is working toward creating an international research center where leading scientists will be engaged in research and development in the fields of anthropomorphic robotics, living systems modeling, and brain and consciousness modeling with the goal of transferring one's individual consciousness to an artificial carrier and achieving cybernetic immortality. 2045 founder Dmitry Itskov's goal is to help humanity with a futuristic reality based on five principles: high spirituality, high culture, high ethics, high science, and high technologies.

In the transhumanist set of ideas, the concept is to keep on growing and learning . . . to redesign, resculpt ourselves. By this, humans could steer evolution by overcoming disease and merging with technology. Because of this, we are breaking out of the predetermined code of our genes and shouting out, "No, I may not want to die just because my genes say it's time to die!"

–Natasha Vita-More, strategic designer, author and innovator within the scientific and technological framework of human enhancement and life extension; interview with Sputnik Futures.

HUMANITY+ IS A NONPROFIT organization dedicated to developing knowledge about the science, technology, and social changes of the twenty-first century. They aim to deeply influence a new generation of thinkers who dare to envision humanity's next steps.

—HumanityPlus.org

The Three Forms of Immortality

We have been oscillating between longevity, life extension, and immortality throughout this journey. We've learned that immortality has yet to be achieved; that Mother Nature has a trick of keeping her youth longer, achieving a biological immortality; that scientists, technologists and transhumanists are working toward life extension and the possibility of a biologically induced "long sleep" to suspend your aging for interstellar travel.

In conclusion, there really is a varied view and definition of what immortality is. Zoltan Istvan, the founder of the Transhumanist Party (yes, there really is one), shared with Sputnik Futures what he believes are the different ways to get to immortality. For example, you can biologically make it so that you can either reverse aging or stop aging through some type of genetic or cellular intervention. Or you could merge with bionic parts, robotic parts, so you're sort of half-cyborg, half-machine. One of the main reasons people die is because of organ failure. So if a bionic heart is available when your heart fails, or if you could have a 3-D bioprinted liver, these are things he believes would be very helpful in keeping you alive, at least until science ultimately cracks the code of extreme longevity. He believes that those of us younger than seventy or eighty years have a very real shot of using technology over the next ten, twenty, or thirty years to live far longer. He thinks that those people who are twenty or twenty-five years of age should have no problem living indefinitely, barring a catastrophe.

But it is the third way that excites Istvan most: "Ultimately, the third form of immortality is the uploading one, mind uploading. And this is the one that I just simply like the best. How far do you take your consciousness expanding? It could be like some enormous, godlike being that just spans the entire universe."

Join the Transhumanist Conversation

Transhumanity.net: A community platform for all things related to transhumanism, featuring hundreds of contributors who aim to help people understand humanity's future now.

Transhumanist Party (US) "is the largest transhumanist political organization in the world—a nonmonetary organization staffed entirely by volunteers and devoted to advocacy of emerging technologies, life extension, and rationality" with the aim of putting science, health, and technology at the forefront of American politics (from Transhumanist-party.org).

Future of Humanity Institute is a multidisciplinary research institute at the University of Oxford with a mission to "shed light on crucial considerations that might shape humanity's long-term future" (see fhi.ox.ac.uk).

2045 Initiative: Immortality to the Rescue

In the 2045 Initiative mission statement, there is a reference to the escalation of interlocking environmental crises "which threaten our planetary habitat and the continued existence of humanity as a species." To forestall this catastrophe and preserve mankind's future, the conceivers of the initiative propose to contribute to the creation of a more safe, sane and sustainable society. The 2045 Initiative claims to devote particular attention to enabling the fullest possible dialogue between the world's major spiritual traditions, science and society.

—2045 Strategic Social Initiative

The great transformation to radical longevity will be a delicate and nuanced balance of humanity and technology. We use tools to improve our lives, and the evolution of our tools will drive our own evolution. A new chapter in humanity's story has begun, and we–living together symbiotically with machine intelligence–get to write what happens next. It's an age in which the marriage of human sensitivity and artificial intelligence fundamentally alters and improves the way we live.

The advances result in humanity surpassing the limits of physical and cognitive capacity. We become part of the Internet of everything, evolving in relationships with automated personalities who are our teachers, companions, and lovers. The lines between technologies that enhance our physical, digital and biological environments have blurred. Where do we go next? To the galaxies, my friend!

Will microgravity unlock
the longevity code?

The space industry is looking beyond the idea of colonizing the moon or Mars, toward the possibility that the zero-gravity landscape of our solar system may hold some keys for targeting aging.

Some of the tech titans investing in the privatization of space, like Elon Musk and Jeff Bezos, are worried that there is no second chance for humanity, so we need to colonize another world. Musk spent his personal fortune on Blue Origin and is focused on colonizing the moon. Others are banking on the medical and health solutions we will find. According to NASA, it seems that some types of stem cells grow faster in simulated microgravity. This is important not only for aging but for the quest to colonize Mars. For humans to live on the moon, Mars, or another other planet, we will need to grow a significant number of human stem cells to treat diseases. Stem cells are vital for survival, as they can divide into all different types of cells; for instance, a heart stem cell can become a liver stem cell. Human stem cells are key to repairing tissue throughout a person's lifetime.

Space is our new research laboratory as scientists look at how human cells can express themselves while in microgravity. The theory is that our cells have memory, and this memory can be triggered in a zero-gravity environment to unlock the right codes for genetic expressions and regeneration.

NASA's Bioculture System is a cell biology research platform for the International Space Station that supports short- and long-duration studies involving the culture of living cells, microbes, and tissues in the unique microgravity environment of space flight.[17] Its experiments look for a deeper understanding of tissue engineering, regeneration, and wound healing—and, perhaps, will potentially unlock the genetic code for longevity, so that one day we can master healthy space travel.

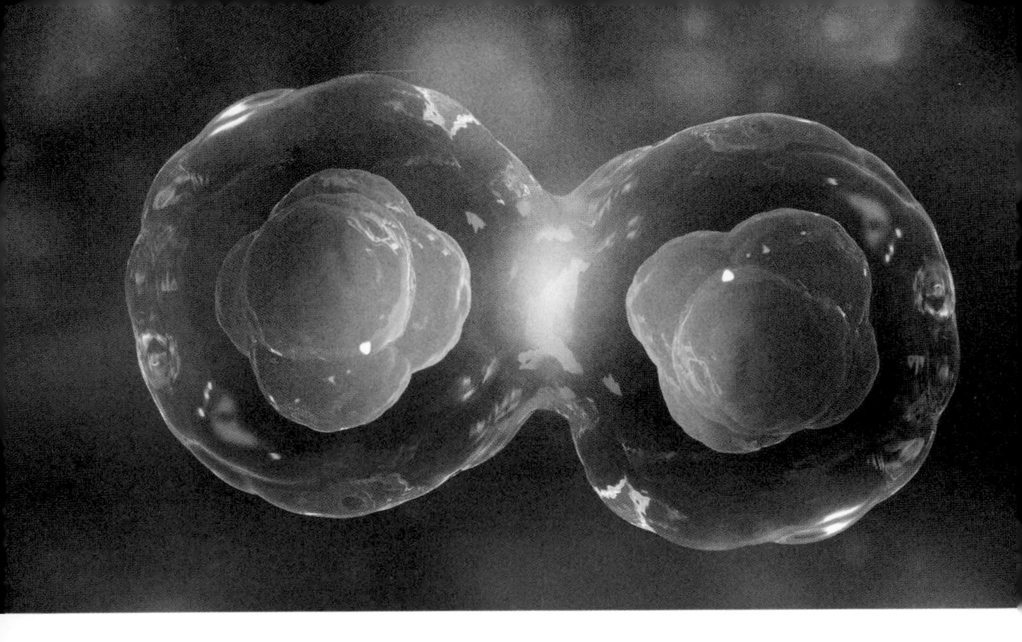

When we do finally journey to Mars and set up new communities on the moon, will these human spacefarers evolve into some other creature?

Robert Bigelow, one of the first pioneers in the privatization of space, has been working on inflatable balloon-like "ships" that can be sent further into the universe. His latest concept, called BB30, is expected to travel to the moon, and according to Bigelow, can accommodate up to four people indefinitely.

When we asked Bigelow in an interview if he thought humans living on a planet other than Earth would eventually speciate, he believed physically we would. But he questioned, "Are we talking about a really spiritual Cro-Magnon creature, or are we talking about something else that physically looks different?" He believes that, yes, we can change, but stressed that the bigger thing to think about is our consciousness and whether it in fact survives death.

There is another way that Dr. Michio Kaku believes we can travel to the far galaxies: through immortal consciousness. He believes that one day we will digitize the entire human brain and shoot it into outer space. In an article on the website Big Think, Dr. Kaku, a renowned theoretical physicist, explains we will do this by putting our "connectome" on a laser and shooting it out to the moon.[18] He is referring to a goal of the Human Connectome Project, which aims to construct a map of the complete structural and functional neural connections in vivo within and across individuals. It is the first large-scale attempt to map the entire brain in about one hundred years. And Dr. Kaku thinks we will shoot it to the stars.

You can travel to space in the morning and be back to Earth that evening. We'll explore the universe as pure consciousness traveling at the speed of light.

–Michio Kaku, theoretical physicist, from the video *How Your Immortal Consciousness Will Travel the Universe*, Big Think, July 2019

Kaku and other experts in the field of human consciousness propose that we may already be immortal through our consciousness. Studies in near-death experiences and the vivid recall of activities that patients see while clinically dead are awakening a new look at the energy that sustains our bodies and our thoughts and the possibility that we are connected to a universal wave. The first law of thermodynamics, also known as the law of conservation of energy, states that energy can neither be created nor destroyed; energy can only be transferred or changed from one form to another. We have an energetic body; our biology runs on molecular energy; our nervous and circulation systems and heartbeats and neural exchanges are all an energetic exchange. If our consciousness is energy, and that energy is already immortal, perhaps we need to set our sights beyond biological reprogramming to devise ways to harness the electrical fields that are within and around us. Ah, hold that thought. More to come in our next book, *Tuning Into Frequency*.

Hacking our humanity for immortality? We are just beginning to script that next chapter in life. Perhaps out there among the stars we will discover the source of our first atoms and the matter that makes us the species we are. After all, we are all stardust (and cue the immortal David Bowie) . . .

Alice in Futureland

A Speculative Life in 2050

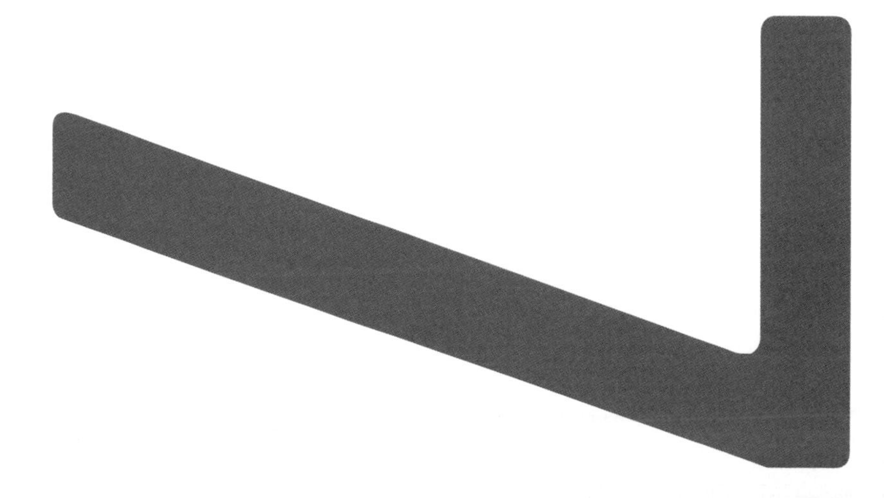

Can we live long enough to live forever?

Does your Instagram feed make you feel #transhuman? Are you popping NAD+ at breakfast? Do you see death as reversible? Phew! Excuse my questions; it's just that I can't surf the net without falling into the longevity craze.

To die or not to die–*that* is the question.

Hello; my name is Alice, and I can live forever. It helps that I am one part human and one part AI. When anyone meets me for the first time, they always ask me: How do you live so long?

And I always say, "*Wonder.*"

I've learned over the years that you have to focus on the positives. I'm an optimist.

Humans have been searching for centuries for the secret to living longer, but the answer may be as simple as maintaining a positive state of mind.

Research suggests that optimistic men and women may live longer than those who are pessimistic. While the association is clear, scientists still do not fully understand why. If you ask me, the AI part of me will say it is the gift of being human. In his first book, *You Are Not a Gadget: A Manifesto*, Silicon Valley visionary Jaron Lanier explains why humans still matter in an age increasingly mediated by computers.[1] I tend to agree. Biology runs deep, and we are only starting to understand our evolutionary code. Mix biology with AI, add a fly named Indy and a few jellyfish, and you have plenty of reasons for optimism.

Have you ever wondered what the future may hold when we can live past 120 healthfully?

Will we all be lining up for some form of human augmentation that supersizes our senses or gives us X-Man abilities? Or will we become what Jean Houston, a leader in the human potential movement, calls the "possible" human?[2] The ability to hack our bodies and our senses will enable us see, hear, touch, taste, and smell things that a mere human, minus AI, could not.

Is our purpose on the planet to understand that we are already immortal? That is, are we spiritual machines? Leading AI advocate Ray Kurzweil has posited that artificial intelligence already has the capability to "feel" and "think" like a human.[3] The twenty-first century will see a blurring of the line between human and machine as neural implants become more prevalent. Eventually, machines will become "spiritual"—or, as Kurzweil conceives it, "conscious."

Mind uploaders are looking for a possible new human through technology. But theoretical physicist John A. Wheeler told us in a Sputnik Futures interview that it is "fantastic that evolution would have ended up with us. What other kind of creature could it have been?" Wheeler was excited about a recent discovery, in China, of dinosaurs that had wings like birds, and he asked, "Do we have something, some faculty that we haven't put to use, the way the dinosaurs had put to use these feathers of theirs until later?"

I wonder!

Follow me for a moment through the looking glass to imagine what new possibilities await us in the next thirty years. Science will make great strides in helping us live longer, healthier, happier lives. Epigenetic therapies, cellular reprogramming, and CRISPR gene editing will most likely become commercialized by 2040, enabling a new approach to medicine that treats the risk of developing a disease as a disease in itself.

Biomarker tests such as epigenetic and DNA methylation will be affordable for everyone. Some people may opt for implants or other forms of augmentation; others may take longevity "vitamins" such as rapamycin, metformin, and NAD+.

It will be the era of truly personalized medicine in which tech-driven therapies deliver precise solutions where and when you need them. It will also be the era of connected health, as AI is expected to proliferate in our everyday lives by 2030, setting the stage for an automated life. Perceptive technology may anticipate our health needs before we even realize we have them.

Finally, we will have the opportunity to be constantly well. If we offset age-related diseases, thereby slowing and perhaps reversing our biological age, we will have extreme generational spread and large populations of supercentenarians. The old infrastructures of government, society, family, business, and education will have to adjust to this new reality. Yes, there are a few unknowns and challenges, but the hope is that as society adjusts, we will make this longer life a happier one.

If only one person at the end reaches immortality, will the experiment of human existence on this planet be successful? Perhaps I will know the answer by the time I'm 120.

Stay tuned!

alice

Alice in Futureland: A speculative and optimistic journey toward an (almost) immortal life*

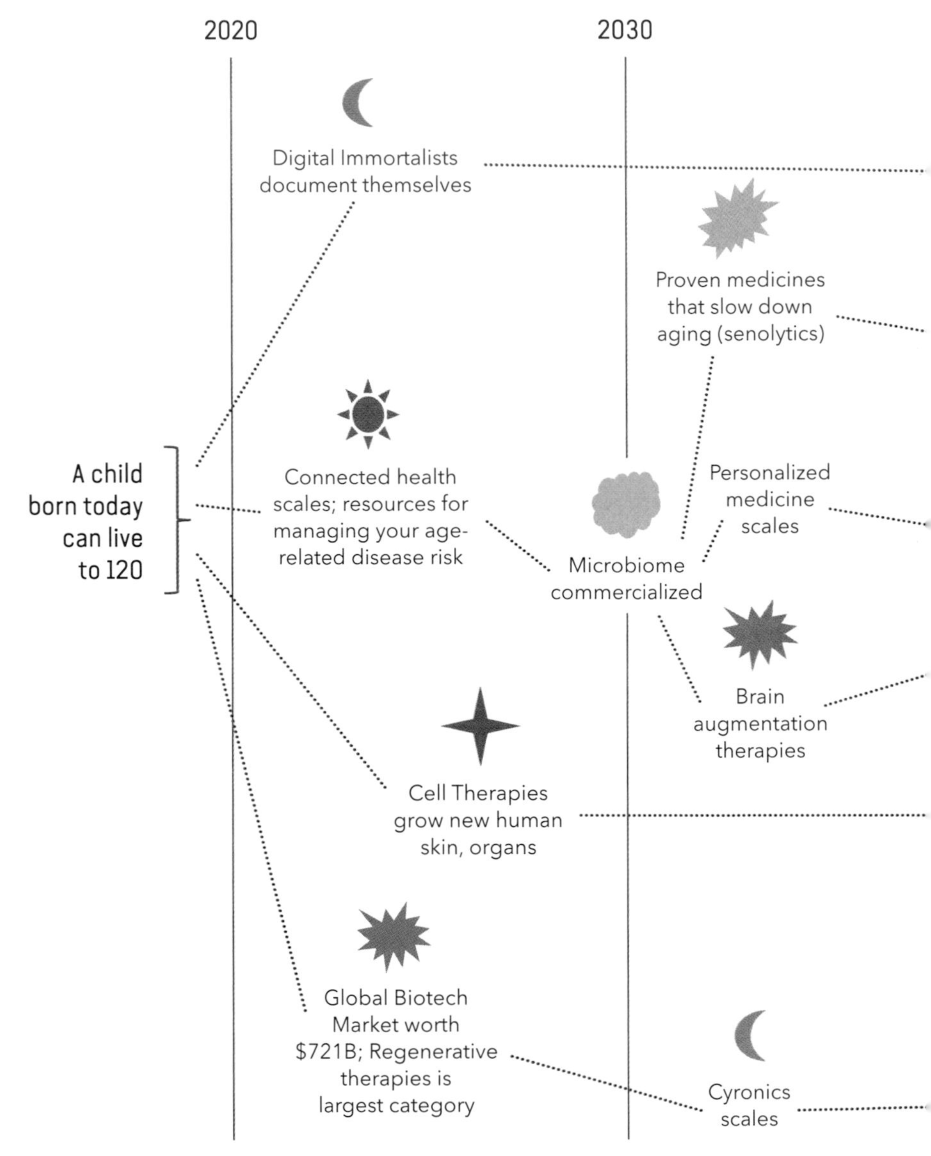

*A speculation, not to be confused with a prediction.

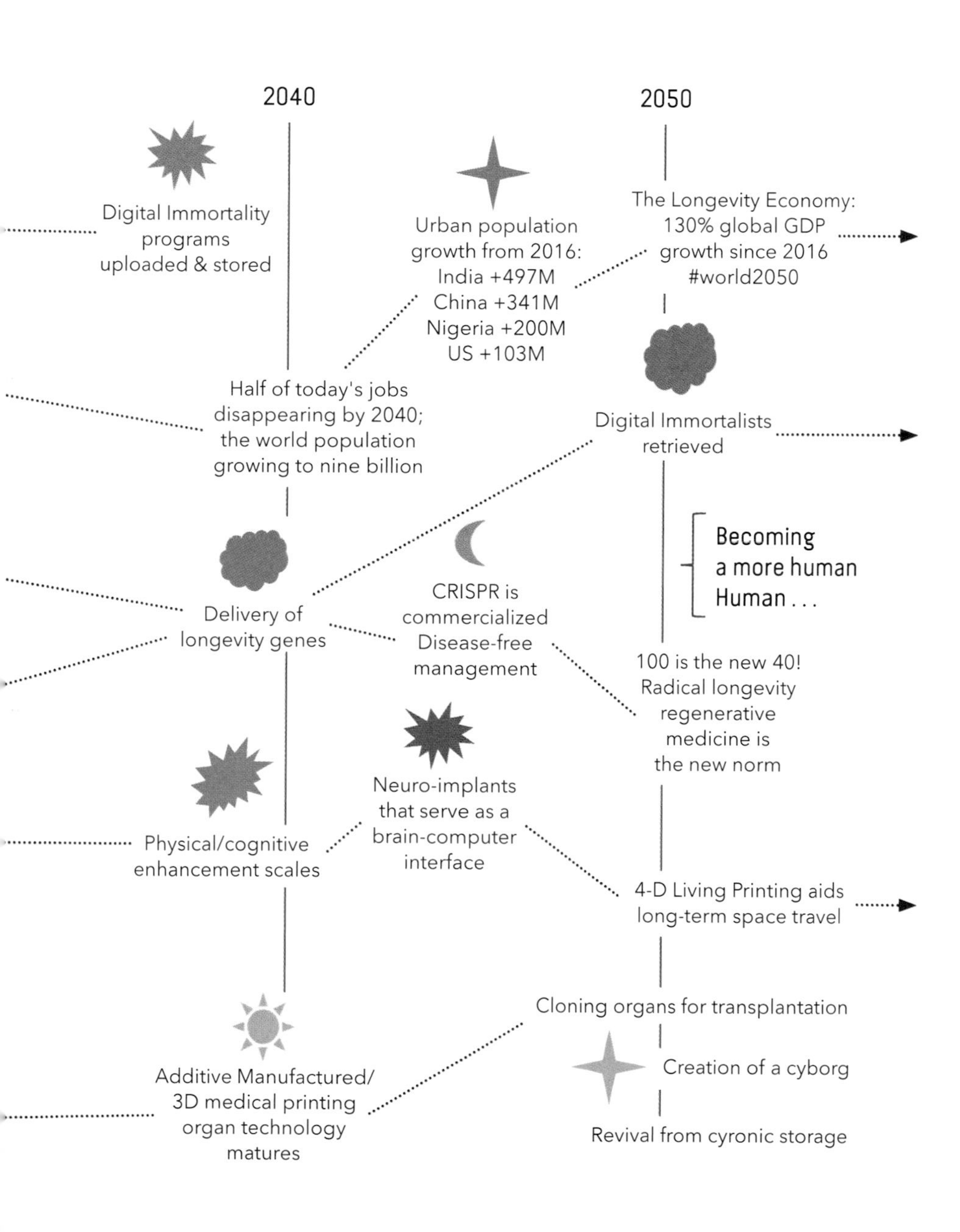
2040
2050
Digital Immortality programs uploaded & stored
Urban population growth from 2016: India +497M China +341M Nigeria +200M US +103M
The Longevity Economy: 130% global GDP growth since 2016 #world2050
Half of today's jobs disappearing by 2040; the world population growing to nine billion
Digital Immortalists retrieved
Becoming a more human Human . . .
Delivery of longevity genes
CRISPR is commercialized Disease-free management
100 is the new 40! Radical longevity regenerative medicine is the new norm
Neuro-implants that serve as a brain-computer interface
Physical/cognitive enhancement scales
4-D Living Printing aids long-term space travel
Cloning organs for transplantation
Creation of a cyborg
Additive Manufactured/ 3D medical printing organ technology matures
Revival from cyronic storage

Down the Rabbit Hole

Articles, Further Reading, People, Organizations

Articles

01: There Is a New Disease Called "Aging": Do You Have It?

Zane Bartlett, "The Hayflick Limit," Embryo Project Encyclopedia, November 14, 2014, https://embryo.asu.edu/pages/hayflick-limit.

Sven Bulterijs, Raphaela S. Hull, Victor C. E. Björk, and Avi G. Roy, "It Is Time to Classify Biological Aging as a Disease," *Frontiers in Genetics* 6 (2015): 205; PMC, June 18, 2015, https://www.frontiersin.org/articles/10.3389/fgene.2015.00205/full.

Cell Press, "How Exercise–Interval Training in Particular–Helps Your Mitochondria Stave Off Old Age," Medical Xpress, March 7, 2017, https://medicalxpress.com/news/2017-03-exerciseinterval-particularhelps-mitochondria-stave-age.html.

Coalition for Radical Life Extension, "Enhance Your Anti-aging Lifestyle," RAADfest. April 26, 2019, https://www.raadfest.com/raadfest/longevity-lifestyle.

"The Complete Guide to the Science of Fasting," Endpoints by Elysium Health, September 15, 2017.

Rose Eveleth, "There Are 37.2 Trillion Cells in Your Body: You Know That Your Body Is Made of Cells–Just How Many? Turns Out That Question Is Not All That Easy to Answer," Smithsonian.com, October 24, 2013, https://www.smithsonianmag.com/smart-news/there-are-372-trillion-cells-in-your-body-4941473/.

Bill Gifford, "Does a Real Anti-aging Pill Already Exist? Inside Novartis's Push to Produce the First Legitimate Anti-Aging Drug," Bloomberg, February 2015.

D. Glick, S. Barth, and K. F. Macleod, "Autophagy: Cellular and Molecular Mechanisms," *Journal of Pathology* 221, no. 1 (May 2010): 3–12; PMC, November 23, 2010, https://www.ncbi.nlm.nih.gov/pmc/articles/PMC2990190/.

Stephen S. Hall, "Finally, the Drug That Keeps You Young," *MIT Technology Review*, October 23, 2018, https://www.technologyreview.com/s/612284/finally-the-drug-that-keeps-you-young/.

Sari Harrar, "Can a Single Pill Keep You Healthy to 100?" *AARP*, July 1,

2019, https://www.aarp.org/health/drugs-supplements/info-2019/pill-drug-aging.html.

Rosamond Hutt, "How Can We Make Healthcare Fit for the Future?" World Economic Forum, January 18, 2016, https://www.weforum.org/agenda/2016/01/how-can-we-make-healthcare-fit-for-the-future/.

Anthony King, "Can We Live Forever?" Chemistry World, January 28, 2019, https://www.chemistryworld.com/features/can-we-live-forever/3009999.article.

Harry McCracken and Lev Grossman, "Can Google Solve Death?" *Time*, September 30, 2013, https://content.time.com/time/subscriber/article/0,33009,2152422,00.html.

Bob Nellis, "Senescent Cell Research Moves into Human Trials," Mayo Clinic News Network, January 7, 2019, https://newsnetwork.mayoclinic.org/discussion/senescent-cell-research-moves-into-human-trials-2/.

S. Jay Olshansky, "From Lifespan to Healthspan," *JAMA* 320, no. 13 (September 2018): 1323–24, https://jamanetwork.com/journals/jama/article-abstract/2703114.

Alvin Powell, "Longevity and Anti-aging Research: 'Prime Time for an Impact on the Globe,'" *Harvard Gazette*, March 8, 2019, https://news.harvard.edu/gazette/story/2019/03/anti-aging-research-prime-time-for-an-impact-on-the-globe/.

Malcolm Ritter, "'Zombie Cells' Buildup in Your Body May Play Role in Aging," Medical Xpress, May 15, 2019, https://medicalxpress.com/news/2019-05-zombie-cells-buildup-body-role.html.

David Sinclair, "This Is Not an Advice Article," LinkedIn Pulse, June 25, 2018, https://www.linkedin.com/pulse/advice-article-david-sinclair.

The State of Aging and Health in America 2013, Centers for Disease Control and Prevention, https://www.cdc.gov/aging/pdf/state-aging-health-in-america-2013.pdf.

A Swiss Scientist Thinks We're in for Trouble: He's Looking for Solutions in Your Mitochondria (podcast), Endpoints by Elysium Health, October 22, 2018, https://www.iheart.com/podcast/-endpoints-by-elysium-health-31087837/episode/a-swiss-scientist-thinks-were-in-33440042/.

"There Will Be No Old People—Anti-Aging Scientist," RT Question More, December 24, 2018, https://www.rt.com/shows/sophieco/447271-anti-aging-health-science/.
"What We Can Do Right Now to Increase Vitality and Reduce Aging Biomarkers," People Unlimited, November 19, 2018, https://peopleunlimitedinc.com/posts/2018/11/19/age-reversal-takeaways-from-raadfest-2018.

02: What's Your GrimAge?

Arizona State University, "Scientists Unveil a Hidden Secret of the Immortality Enzyme Telomerase," Medical Xpress, February 23, 2018, https://asunow.asu.edu/20180223-discoveries-asu-scientists-unveil-immortality-enzyme-telomerase.
Christopher Bergland, "Emotional Distress Can Speed Up Cellular Aging: Chronic Stress Accelerates Premature Aging by Shortening DNA Telomeres," *Psychology Today*, April 7, 2014, https://www.psychologytoday.com/us/blog/the-athletes-way/201404/emotional-distress-can-speed-cellular-aging.
Yuxia Cui, David M. Balshaw, Richard K. Kwok, Claudia L. Thompson, Gwen W. Collman, and Linda S. Birnbaum, "The Exposome: Embracing the Complexity for Discovery in Environmental Health," *Environmental Health Perspectives* 124, no. 8 (August 2016): A137–A140; PMC, August 1, 2016, https://www.ncbi.nlm.nih.gov/pmc/articles/PMC4977033/##targetText=The%20exposome%2C%20which%20is%20defined,discovery%20in%20environmental%20health%20research.
K. K. Dennis, S. S. Auerbach, D. M. Balshaw, Y. Cui, M. D. Fallin, M. T. Smith, A. Spira, S. Sumner, and G. W. Miller, "The Importance of the Biological Impact of Exposure to the Concept of the Exposome," *Environmental Health Perspectives* 124, no. 10 (October 2016): 1504–1510; PMC, June 3, 2016, https://www.ncbi.nlm.nih.gov/pmc/articles/PMC5047763/.
Oren Froy, "Circadian Rhythms, Aging and Life Span in Mammals," *Physiology*, August 1, 2011, https://www.physiology.org/doi/full/10.1152/physiol.00012.2011.

Fedor Galkin, Aleksandr Aliper, Evgeny Putin, Igor Kuznetsov, Vadim M. Gladyshev, and Alex Zhavoronkov, "Human Microbiome Aging Clocks Based on Deep Learning and Tandem of Permutation Feature Importance and Accumulated Local Effects," bioRxiv: The Preprint Server for Biology, December 28, 2018, https://www.biorxiv.org/content/10.1101/507780v1.full.

Eni Gómez, Eva Toribio, Montserrat Delor, and Judit Morlà Folch, "Hair Exposome, a Concept to Address Hair Disorders from a New Perspective by Beauty Cluster Barcelona," Skinobs Cosmetic Testing News, February 28, 2018, https://skinobs.com/news/en/non-classe-en/hair-exposome-a-concept-to-address-hair-disorders-from-a-new-perspective-by-beauty-cluster/.

Cother Hajat and Emma Stein, "The Global Burden of Multiple Chronic Conditions: A Narrative Review," Preventive Medicine Reports 12 (December 2018): 284–93; PMC, October 19, 2018, https://www.ncbi.nlm.nih.gov/pmc/articles/PMC6214883/.

Erika Hayasaki, "Has This Scientist Finally Found the Fountain of Youth?" *MIT Technology Review*, August 8, 2019, https://www.technologyreview.com/s/614074/scientist-fountain-of-youth-epigenome/.

Steve Horvath, "DNA Methylation Age Calculator," https://dnamage.genetics.ucla.edu/home.

Steve Horvath, "DNA Methylation Age of Human Tissues and Cell Types," *Genome Biology* 14, no. 10 (2013): R115, https://genomebiology.biomedcentral.com/articles/10.1186/gb-2013-14-10-r115.

Efraim Jaul, and Jeremy Barron, "Age-Related Diseases and Clinical and Public Health Implications for the 85 Years Old and Over Population," *Frontiers in Public Health* 5 (December 11, 2017): 335; PMC, December 11, 2017, https://www.ncbi.nlm.nih.gov/pmc/articles/PMC5732407/.

Kyoung-Nam Kim and Yun-Chu Hong, "The Exposome and the Future of Epidemiology: A Vision and Prospect," *Environmental Health and Toxicology* 32 (May 3, 2017); PMC, May 3, 2017, https://www.ncbi.nlm.nih.gov/pmc/articles/PMC5575673/.

Carlos López-Otin, Maria A. Blasco, Linda Partridge, Manuel Serrano, and Guido Kroemer; "The Hallmarks of Aging," *Cell* 153, no. 6 (June 6, 2013): 1194–1217, https://www.cell.com/fulltext/S0092-8674(13)00645-4.

Terra Marquette, "'Longevity Gene' Protein Offers Possibility of Extended, Healthier Lives," StudyFinds, July 25, 2019, https://www.studyfinds.org/longevity-gene-protein-offers-possibility-extended-healthier-lives/##targetText=Italian%20researchers%20behind%20the%20study,linked%20to%20a%20longer%20life.

National Institutes of Health, "What Is a Gene Mutation, and How Do Mutations Occur?" November 26, 2019.

Alice Park, "We're One Step Closer to a Blood Test That Predicts When a Person Will Die," *Time*, August 21, 2019, https://time.com/5656767/blood-test-longevity/?utm_source =newsletter&utm_medium=email&utm_campaign=the-brief&utm_content=20190831&xid=newsletter-brief.

Brian Parker, "Circadian Rhythm, Sleep, and Aging," *Life Extension Magazine*, 2018, https://www.lifeextension.com/magazine/2018/ss/circadian-rhythm-sleep-and-aging.

"Promising New Biomarker to Retrospectively Capture the Early-Life Exposome," Human Exposome Project, https://humanexposomeproject.com/news/promising-new-biomarker-to-retrospectively-capture-the-early-life-exposome/.

Robert Sanders, "Long-Sought Structure of Telomerase Paves Way for Drugs for Aging, Cancer," Berkeley News, April 25, 2018, https://news.berkeley.edu/2018/04/25/long-sought-structure-of-telomerase-paves-way-for-drugs-for-aging-cancer/.

Chris Sweeney, "Uncovering a 'Smoking Gun' of Biological Aging," Harvard T.H. Chan School of Public Health, February 14, 2019, https://www.hsph.harvard.edu/news/press-releases/rdna-biological-aging-clock/.

Helen Thomson, "Drug Cocktail Seems to Reverse Biological Signs of Ageing in People," New Scientist, September 5, 2019, https://www.newscientist.com/article/2215537-drug-cocktail-seems-to-reverse-biological-signs-of-ageing-in-people/.

Umeå University, "Depression and Chronic Stress Accelerates Aging,"

ScienceDaily, November 10, 2011, https://www.sciencedaily.com/releases/2011/11/111109093729.htm.
Kate Whitehead, "How to Look Younger Than You Are: Biomarkers–Which to Track and How to Step In If Needed," *South China Morning Post*, August 5, 2019, https://www.scmp.com/lifestyle/health-wellness/article/3021225/how-look-younger-you-are-bio markers-which-track-and-how.
Tina Woods, "'Longevity' Could Reach Billions in 2019—and Is No Longer Just the Preserve of Billionaires," *Forbes*, January 11, 2019, https://www.forbes.com/sites/tinawoods/2019/01/11/longevity-could-reach-billions-in-2019-and-is-no-longer-just-the-preserve-of-billionaires/#5a21565c668f.
Colleen Zacharyczuk, "Human Microbiome Project May Hold Promise for the Future," Healio, June 2012, https://www.healio.com/pediatrics/news/print/infectious-diseases-in-children/%7B39d199b9-64c6-4345-90fa-f96293fb258a%7D/human-microbiome-project-may-hold-promise-for-future.

03: Is Nature Immortal?

"Aging Gracefully with the Help of Marmosets," Texas Biomedical Research Institute News, April 2, 2019, https://www.txbiomed.org/news-press/news-releases/aging-gracefully-with-the-help-of-marmosets/##targetText=%E2%80%9CFor%20these%20reasons%2C%20the%20marmoset,authors%20of%20the%20study%20write.
Matthew Blakeslee, "Long-Lived Yeast May Hold a Key to Human Longevity," USC News, April 20, 2001, https://news.usc.edu/5375/Long-lived-yeast-may-hold-a-key-to-human-longevity/.
David Brin, "Do We Really Want Immortality?," Institute for Ethics and Emerging Technologies, January 8, 2012, https://ieet.org/index.php/IEET2/print/5068.
Christian-Albrechts-Universität zu Kiel, "Solving the Mystery of Aging: Longevity Gene Makes Hydra Immortal and Humans Grow Older," *ScienceDaily*, November 13, 2012, https://www.sciencedaily.com/releases/2012/11/121113091953.htm.

Heidi Gardner, "Can CRISPR Stop or Reverse Aging?" Synthego, July 3, 2018, https://www.synthego.com/blog/can-crispr-stop-us-from-aging.

Tessa Gregory, "Tardigrades' Genes Help Them Survive Extreme Conditions," *PLoS Research News*, July 28, 2017, https://researchnews.plos.org/2017/07/28/research-round-up-tardigrades-genes-help-them-survive-extreme-conditions-women-and-children-also-exposed-to-hunting-related-pathogens-isotopes-in-neolithic-cattle-teeth-suggest-a-variety-o/.

Stephen S. Hall, "Finally, the Drug That Keeps You Young, *MIT Technology Review*, October 23, 2018, https://www.technologyreview.com/s/612284/finally-the-drug-that-keeps-you-young/.

Charlotte Hu, "Animals That Defy the Rules of Aging–Like Naked Mole Rats–Could Help Scientists Unravel the Secrets to Longevity," Business Insider, August 15, 2018, https://www.businessinsider.sg/animals-that-defy-aging-rules-offer-longevity-clues-2018-8/.

Charlotte Hu, "A Scientist Who Studies Aging Reveals How Restricting Calories Might Offer Protection Against Age-Related Diseases," Business Insider, August 23, 2018, https://www.businessinsider.com/restricting-calories-could-protect-against-aging-2018-8.

Kaine Korzekwa, "Radiation-Resistant E. coli Evolved in the Lab Give View into DNA Repair," University of Wisconsin–Madison News, February 26, 2019, https://news.wisc.edu/radiation-resistant-e-coli-evolved-in-the-lab-give-view-into-dna-repair/.

Anita Krisko and Miroslav Radman, "Biology of Extreme Radiation Resistance: The Way of *Deinococcus radiodurans*," *Cold Spring Harbor Perspectives in Biology* 5, no. 7 (July 2013): a012765; PMC, July 2013, https://cshperspectives.cshlp.org/content/5/7/a012765.full.

Rute Martins, Gordon J. Lithgow, and Wolfgang Link, "Long Live FOXO: Unraveling the Role of FOXO Proteins in Aging and Longevity," *Aging Cell* 15, no. 2 (April 2016): 196–207; PMC, December 8, 2015, https://www.ncbi.nlm.nih.gov/pmc/articles/PMC4783344/.

Daniel Oberhaus, "A Crashed Israeli Lunar Lander Spilled Tardigrades on the Moon," *Wired*, August 5, 2019, https://www.wired.com/story/a-crashed-israeli-lunar-lander-spilled-tardigrades-on-the-moon/.

Reason, "Animal Data Shows Fisetin to be a Surprisingly Effective Senolytic," Fight Aging, October 3, 2018, https://www.fightaging.org/archives/2018/10/animal-data-shows-fisetin-to-be-a-surprisingly-effective-senolytic/.

John Rennie and Lucy Reading-Ikkanda, "Seeing the Beautiful Intelligence of Microbes," *Quanta Magazine*, November 13, 2017, https://www.quantamagazine.org/the-beautiful-intelligence-of-bacteria-and-other-microbes-20171113/.

Salk Institute, "CRISPR/Cas9 Therapy Can Suppress Aging, Enhance Health and Extend Life Span in Mice," *ScienceDaily*, February 19, 2019, https://www.sciencedaily.com/releases/2019/02/190219111747.htm.

Alexandra Sifferlin, "See 10 Animals That Live a Ridiculously Long Time," *Time*, February 11, 2016, https://time.com/4216390/animals-longevity/.

What Can Scientists Learn from Animals That Defy "Normal Aging"? (podcast), Endpoints Science Publication by Elysium Health, September 2018, https://open.spotify.com/episode/5z73IbOBMFY2ilwMrrbt4e.

Ed Yong, "Tardigrades Become First Animals to Survive Vacuum of Space," *National Geographic*, September 8, 2008, https://www.nationalgeographic.com/science/phenomena/2008/09/08/tardigrades-become-first-animals-to-survive-vacuum-of-space/##targetText=By%20Ed%20Yong&targetText=In%20September%20last%20year%2C%20a,to%20the%20vacuum%20of%20space.

04: Will One Hundred Be the New Forty?

Dave Asprey, "13 Anti-aging Supplements to Turn You into Benjamin Button," Bulletproof, https://www.bulletproof.com/supplements/age-immune/anti-aging-supplements/.

"The Baby Bonanza," *Economist*, August 27, 2009, https://www.economist.com/briefing/2009/08/27/the-baby-bonanza.

Lewis Braham, "Forget Your Real Age, Plan Your Retirement around Your 'Biological Age,'" *Barron's*, January 27, 2019, https://www.barrons.com/articles/retirement-biological-age-51548356662.
Kaare Christensen, Gabriele Doblhammer, Roland Rau, and James W. Vaupel, "Ageing Populations: The Challenges Ahead," *Lancet* 374, no. 9696 (October 3, 2009): 1196–1208; PMC, January 25, 2010.
The Demographic Dividend: Investing in Human Capital, website hosted by the Bill & Melinda Gates Institute for Population and Reproductive Health at the John Hopkins Bloomberg School of Public Health, http://www.demographicdividend.org/.
"Designing for Immortality: Reversible Destiny Lofts/Gins & Arakawa," Directions for Use of the Reversible Destiny Lofts, ArchEyes, January 6, 2016, http://archeyes.com/reversible-destiny-lofts-madeline-gins-and-shusaku-arakawa/.
Alessandro Gandolfi/Parallelozero, "The Village of Longevity," Story Institute, http://www.thestoryinstitute.com/ogimi.
Peter Martin, Grace da Rosa, Ilene C. Siegler, Adam Davey, Maurice MacDonald, and Leonard W. Poon, "Personality and Longevity: Findings from the Georgia Centenarian Study," *Age* 28, no. 4 (December 2006): 343–52; PMC, November 21, 2006, https://www.ncbi.nlm.nih.gov/pmc/articles/PMC3259159/##targetText=Results%20of%20a%20personality%20study,younger%20groups%20(Martin%202002).
"New England Centenarian Study," Boston University School of Medicine, https://www.bumc.bu.edu/centenarian/.
"The Power 9® Evidence-Based Common Lifestyle Denominators Among All Blue Zones," Blue Zones.
"Research: FOXO3 'Longevity Gene' Provides Huge Protection Against Death from Heart Attack," John A. Burns School of Medicine, University of Hawai'i at Mānoa, April 12, 2016, https://jabsom.hawaii.edu/research-foxo3-longevity-gene-provides-huge-protection-against-against-death-from-heart-attack/.
Meera Lee Sethi, "100 Is the New 65," Greater Good Magazine, October 1, 2009, https://greatergood.berkeley.edu/article/item/100_is_the_new_65.

UN Aging Projections, "World's Population Increasingly Urban with More Than Half Living in Urban Areas," United Nations, July 10, 2014, https://www.un.org/en/development/desa/news/population/world-urbanization-prospects-2014.html##targetText=World's%20population%20increasingly%20urban%20with%20more%20than%20half%20living%20in%20urban%20areas,-10%20July%202014&targetText=Today%2C%2054%20per%20cent%20of,66%20per%20cent%20by%202050.&targetText=By%202050%2C%20India%20is%20projected,million%20and%20Nigeria%20212%20million.

Christopher Wareham, "How Can Life-Extending Treatments Be Available for All?" Aeon, August 2, 2017, https://aeon.co/ideas/how-can-life-extending-treatments-be-available-for-all.

Robert Young and John M. "Johnny" Adams, "GRC World Supercentenarian Rankings List," Gerontology Research Group, September 26, 2019.

05: Will We Become Software?

Ade Adeniji, "Not Just Sci-Fi: A Tech Couple Works to Achieve Immortality," Inside Philanthropy, October 25, 2018, https://www.insidephilanthropy.com/home/2018/10/25/not-just-sci-fi-a-tech-couple-works-to-achieve-immortality.

Robert D. Austin and Gary P. Pisano, "Neurodiversity as a Competitive Advantage," *Harvard Business Review*, May–June 2017, https://www.hbs.edu/faculty/Pages/item.aspx?num=52624.

Anthony Cuthbertson, "Elon Musk's Neuralink Plans to Hook Human Brains Directly to Computers," *Independent*, July 17, 2019, https://www.independent.co.uk/life-style/gadgets-and-tech/news/elon-musk-neuralink-brain-computer-transhumanism-threads-a9007756.html##targetText=Entrepreneur%20says%20brain%20chip%20will,of%20symbiosis%20with%20artificial%20intelligence'&targetText=Elon%20Musk's%20secretive%20neurotechnology,brain%20directly%20to%20a%20computer.

"Digital Remains Should Be Treated Like Physical Ones," University of Oxford, April 18, 2018, http://www.ox.ac.uk/news/2018-04-18-digital-remains-should-be-treated-physical-ones.

Tim Fiori, "DARPA and the Brain Initiative," *Medium*, https://medium.com/hummtech/darpa-and-the-brain-initiative-5fa45e9557a.
Alastair Himmer, "Kyoto Temple Puts Faith in Robot Priest," *Japan Times*, August 15, 2019, https://www.japantimes.co.jp/news/2019/08/15/business/tech/kyoto-temple-robot-priest/.
Courtney Humphries, "Digital Immortality: How Your Life's Data Means a Version of You Can Live Forever," *MIT Technology Review*, October 18, 2018, https://www.technologyreview.com/s/612257/digital-version-after-death/.
"Imagining a New Interface: Hands-Free Communication Without Saying a Word," Tech@Facebook, July 30, 2019, https://tech.fb.com/imagining-a-new-interface-hands-free-communication-without-saying-a-word/.
Sharon Kirkey, "Life—After Life: Does Consciousness Continue after Our Brains Die?" *National Post*, April 18, 2019, https://nationalpost.com/news/canada/life-after-life-does-consciousness-continue-after-our-brain-dies.
Clifford N. Lazarus, "Does Consciousness Exist Outside of the Brain?" *Psychology Today*, June 26, 2019, https://www.psychologytoday.com/us/blog/think-well/201906/does-consciousness-exist-outside-the-brain.
Markus Mooslechner, "Supersapiens, the Rise of the Mind," Kurzweil AI, July 21, 2017, https://www.kurzweilai.net/supersapiens-the-rise-of-the-mind.
Antonia Regalado, "A Startup Is Pitching a Mind-Uploading Service That Is '100 Percent Fatal,'" *MIT Technology Review*, March 13, 2018, https://www.technologyreview.com/s/610456/a-startup-is-pitching-a-mind-uploading-service-that-is-100-percent-fatal/.
Scott Seckel, "In the Future, You Will Be Forever," Arizona State University, Center for the Study of Religion and Conflict, August 3, 2018, https://csrc.asu.edu/content/future-you-will-be-forever.
"21 Neurotech Startups to Watch: Brain-Machine Interfaces, Implantables, and Neuroprosthetics," CB Insights, January 28, 2019, https://www.cbinsights.com/research/neurotech-startups-to-watch/.
Carolyn Wilke, "Can Humans Sense Magnetic Fields?" *Scientist*,

March 19, 2019, https://www.the-scientist.com/news-opinion/can-humans-sense-the-magnetic-field--65611.

Caroline Winter, "Upgrade Your Memory with a Surgically Implanted Chip," Bloomberg Businessweek, June 10, 2019, https://www.bloomberg.com/news/articles/2019-06-10/upgrade-your-memory-with-a-surgically-implanted-chip.

06: What Will It Mean to Be Human?

Nicholas A. Christakis, "How AI Will Rewire Us," *Atlantic*, April 2019, https://www.theatlantic.com/magazine/archive/2019/04/robots-human-relationships/583204/.

Dr. Al Emondi, "Next-Generation Nonsurgical Neurotechnology," DARPA, https://www.darpa.mil/program/next-generation-nonsurgical-neurotechnology.

J. H. Fowler and N. A. Christakis, "Dynamic Spread of Happiness in a Large Social Network: Longitudinal Analysis over 20 Years in the Framingham Heart Study," *British Medical Journal* 337 (December 5, 2008): a2338.

Kristen French, "Does Your Mind Control How You Age?" Endpoints/Elysium Health, July 19.

Kristen French and Lola Dupre, "Anger Is Normal. But Scientists Say the Toxic Kind Can Shorten Your Life," Endpoints/Elysium Health, August 26.

Jennifer E. Graham, Lisa M. Christian, and Janice K. Kiecolt-Glaser, "Stress, Age, and Immune Function: Toward a Lifespan Approach," *Journal of Behavioral Medicine* 29, no. 4 (August 2006): 389–400; PMC, May 19, 2006, https://www.ncbi.nlm.nih.gov/pmc/articles/PMC2805089/.

Michio Kaku, *How Your Immortal Consciousness Will Travel the Universe* (video and transcript), Big Think, July 29, 2019, https://bigthink.com/videos/michio-kaku-2639431116.

Becca R. Levy, Martin D. Slade, Stanislav V. Kasl, and Suzanne R. Kunkel, "Longevity Increased by Positive Self-Perceptions of Aging," *Journal of Personality and Social Psychology* 83, no. 2 (August 2002): 261–70, https://www.apa.org/pubs/journals/releases/psp-832261.pdf.

Becca R. Levy, Martin D. Slade, Robert H. Pietrzak, and Luigi Ferrucci, "Positive Age Beliefs Protect Against Dementia Even Among Elders with High-Risk Gene," *PLoS ONE*, February 7, 2018, https://journals.plos.org/plosone/article?id=10.1371/journal.pone.0191004.

"Optimists Live Longer: New Study by BU, Harvard Researchers Indicates Positive Thinking Can Help Us Live Past Age 85," Brink: Pioneering Research from Boston University, August 26, 2019, http://www.bu.edu/articles/2019/optimists-live-longer/.

Mike Perry, "The First Suspension," Alcor Life Extension Foundation; article from the column "For the Record," *Cryonics*, July 1991, https://alcor.org/Library/html/BedfordSuspension.html.

Michael A. Rose, "55 Theses on the Power and Efficacy of Natural Selection for Sustaining Health," 55Theses.org, https://55theses.org/the-55-theses/.

Michael Schermer, "The Immortalists–Can Science Defeat Death?" Science Focus, July 1, 2019, https://www.sciencefocus.com/future-technology/the-immortalists-can-science-defeat-death/.

Theødor, "Natasha Vita-More Doesn't Want to Play God. She Just Wants to Play," *Medium*, September 2, 2017, https://medium.com/tryangle-magazine/natasha-vita-more-doesnt-want-to-play-god-she-just-wants-to-play-611cafe85d6.

Additional Reading

Asprey, Dave, *The Bulletproof Diet: Lose Up to a Pound a Day, Reclaim Energy and Focus, Upgrade Your Life* (New York: Rodale Books, 2014; reprint edition 2018).

Blackburn, Elizabeth, and Elissa Epel, *The Telomere Effect: A Revolutionary Approach to Living Younger, Healthier, Longer* (New York: Grand Central, 2017; reprint edition 2018).

Blueprint: The Evolutionary Origins of a Good Society by Nicholas A. Christakis, YaleNews book review, March 26, 2019, https://news.yale.edu/2019/03/26/blueprint-evolutionary-origins-good-society.

Buettner, Dan, *The Blue Zones: 9 Lessons for Living Longer from the*

People Who've Lived the Longest (Washington, DC: National Geographic, 2008; second edition 2012).
Church, George M., and Ed Regis, *Regenesis: How Synthetic Biology Will Reinvent Nature and Ourselves* (New York: Basic Books, 2014).
Ettinger, Robert, *The Prospect of Immortality* (New York: Doubleday, 1964).
Istvan, Zoltan, *The Transhumanist Wager* (e-book; Futurity Imagine Media, 2013).
Keltner, Dacher, *Born to Be Good: The Science of a Meaningful Life* (New York: W. W. Norton, 2009).
Lu, Timothy, "Synthetic Biology to Reprogram Life," *Live Long and Master Aging* podcast, January 22, 2019, http://www.llamapodcast.com/timothy-lu/##targetText=Timothy%20Lu%3A%20Synthetic%20Biology%20Group%2C%20MIT&targetText=The%20idea%20is%20that%20capsules,the%20need%20for%20a%20colonoscopy.
Milevsky, Moshe A., *Longevity Insurance for a Biological Age* (independently published, 2019).
Rothblatt, Martine, *Virtually Human: The Promise—and the Peril—of Digital Immortality*; foreword by Ray Kurzweil (New York: Picador, 2015).
Sinclair, David A., with Matthew D. LaPlante, *Lifespan: Why We Age—and Why We Don't Have To* (New York: Atria Books, 2019).
Willcox, Bradley J., D. Craig Willcox, and Makoto Suzuki, *The Okinawa Diet Plan: Get Leaner, Live Longer, and Never Feel Hungry* (New York: Three Rivers Press, 2004; Harmony reprint 2005).
Willcox, Bradley J., D. Craig Willcox, and Makoto Suzuki, *The Okinawa Program: How the World's Longest-Lived People Achieve Everlasting Health—and How You Can Too* (New York: Three Rivers Press, 2001; Harmony reprint 2002).

People We Noted

Dave Asprey, CEO and founder of Bulletproof; bulletproof.com
Johan Auwerx, MD, PhD, professor at the École Polytechnique Fédé-

rale in Lausanne, Switzerland, where he directs the Laboratory for Integrated and Systems Physiology (LISP); ahlresearch.org

Ilaria Bellantuono, professor, Department of Oncology and Metabolism, and codirector of the Healthy Lifespan Institute, University of Sheffield; sheffield.ac.uk/oncology-metabolism

Juan Carlos Izpisúa Belmonte, professor, Gene Expression Laboratory, Salk Institute; salk.edu

Robert Bigelow, founder and president of Bigelow Aerospace; bigelowaerospace.com

Elizabeth Blackburn and Carol Greider (2009 Nobel Prize in Physiology or Medicine); nobelprize.org

Mikhail Blagosklonny, PhD, MD, Roswell Park Comprehensive Cancer Center; roswellpark.org

Judith Campisi, PhD, professor at the Buck Institute for Research on Aging; senior Scientist at Lawrence Berkeley National Laboratory; and cofounder of Unity Biotechnology; unitybiotechnology.com

Julian Chen, professor, School of Molecular Sciences at Arizona State University; sms.asu.edu

Nicholas A. Christakis, MD, PhD, MPH, Sterling Professor of Social and Natural Science at Yale University; director of the Human Nature Lab; codirector of the Yale Institute for Network Science; human naturelab.net

George Church, PhD, American geneticist, molecular engineer, and chemist; founder of the Human Genome Project; director of PersonalGenomes.org; wyss.harvard.edu

Erik Davis, author, podcaster, award-winning journalist, and popular speaker; techgnosis.com

Paul Ekman, PhD, the world's foremost expert in facial expressions; professor emeritus at the University of California Medical School in San Francisco; paulekman.com

Walter Greenleaf, behavioral neuroscientist and medical technology developer, Stanford University; mediax.stanford.edu

Leonard Guarente, Novartis Professor of Biology, MIT/The Leonard Guarente Lab at MIT; web.mit.edu

Steve Horvath, professor of human genetics and biostatistics, UCLA Fielding School of Public Health; ph.ucla.edu

Masahiko Inami, professor of information somatics, Research Center for Advanced Science and Technology, University of Tokyo; rcast .u-tokyo.ac.jp

Hiroshi Ishiguro, director of the Intelligent Robotics Laboratory, Graduate School of Engineering Science, Osaka University, Japan; robotics pioneer; eng.irl.sys.es.osaka-u.ac.jp

Zoltan Istvan, founder, Transhumanist Party; zoltanistvan.com

Dmitry Itskov, 2045 Initiative; 2045.com

Naveen Jain, founder of companies including Moon Express, Viome, BlueDot, TalentWise, Intelius, and InfoSpace; naveenjain.com

Bryan Johnson, founder of Kernel, OS FUND, and Braintree; bryan johnson.co

Matt R. Kaeberlein, PhD, pathology professor at the Washington Medical Center in Seattle; codirector, University of Washington Nathan Shock Center of Excellence in the Basic Biology of Aging; director, Healthy Aging and Longevity Research Institute; president, American Aging Association; pathology.washington.edu

Dacher Keltner, professor of psychology at University of California, Berkeley, and founding director of the Greater Good Science Center; psychology.berkeley.edu

Valter Longo, director of the Longevity Institute at USC Leonard Davis School of Gerontology; caloric restriction researcher; inventor of the ProLon Fasting Mimicking Diet; valterlongo.com

Timothy K. Lu, professor of biological engineering, electrical engineering, and computer science, and core faculty member, Synthetic Biology Center, MIT; mit.edu

Moshe A. Milevsky, leading authority on the intersection of wealth management, financial mathematics, and insurance; moshemi levsky.com

S. Jay Olhansky, epidemiologist; professor, School of Public Health, University of Illinois at Chicago; sjayolshansky.com

Michael R. Rose, professor of ecology and evolutionary biology, University of California, Irvine; faculty.uci.edu

David Sinclair, professor in the Department of Genetics at Harvard Medical School and codirector of the Paul F. Glenn Center for the Biological Mechanisms of Aging; genetics.med.harvard.edu

Esther Sternberg, professor of medicine in the University of Arizona College of Medicine; research director of the Arizona Center for Integrative Medicine; and director of the University of Arizona Institute on Place and Wellbeing; medicine.arizona.edu

Natasha Vita-More, PhD, author and coeditor of the Transhumanist Reader; designer of Primo Posthuman; international speaker on human enhancement, radical life extension, and humanity's future; chair of Humanity+; member of the faculty of the University of Advancing Technology, Tempe, Arizona; natashavita-more.com

Shinya Yamanaka, winner of the Nobel Prize in Physiology or Medicine; nobelprize.org/prizes

Organizations

2045 Initiative; www.2045.com
Academy for Health and Lifespan Research; www.ahlresearch.org
Alcor Life Extension Foundation; www.alcor.org
Amino Labs; www.amino.bio
Apeiron Center; www.apeironcenter.com
Biocybernaut Institute; www.biocybernaut.com
Bioquark; www.bioquark.com
Brain Backups; www.brainbackups.com
BrainCo; www.brainco.tech
Bulletproof; www.bulletproof.com
Calico; www.calicolabs.com
Carboncopies; www.carboncopies.org
Celularity; www.celularity.com
Center for PostNatural History; www.postnatural.org
ChromaDex; www.chromadex.com
Coalition for Radical Life Extension; www.rlecoalition.com
Elysium Health; www.elysiumhealth.com
Eternime; www.eterni.me
FitBiomics; www.fitbiomics.com
Future of Humanity Institute, University of Oxford; www.fhi.ox.ac.uk
GenuCure; www.genucurelabs.com

Greater Good Science Center; www.ggsc.berkeley.edu
Health Nucleus; www.healthnucleus.com
Human Ageing Genomic Resources; www.genomics.senescence.info
Human Longevity, Inc.; www.humanlongevity.com
Humanity+; www.humanityplus.org
I Am Human (film); www.iamhumanfilm.com
Insilico Medicine; www.insilico.com
Kernel; www.kernel.co
Life Biosciences; www.lifebiosciences.com
LifeNaut; www.lifenaut.com
Medical Avatar; www.medicalavatar.com
Methuselah Foundation; www.mfoundation.org
myDNAge; www.mydnage.com
Neuralink; www.neuralink.com
Neuroscape; www.neuroscape.ucsf.edu
NIH Human Microbiome Project; www.hmpdacc.org
Novartis; www.novartis.com
OISIN Biotechnologies; www.oisinbio.com
Okinawa Research Center for Longevity Science; www.orcls.org
Openwater; www.openwater.cc
OS Fund; www.osfund.co
People Unlimited; www.peopleunlimitedinc.com
ProLon; www.prolonpro.com
PureTech; www.puretechhealth.com
RAADfest; www.raadfest.com
RealAge; www.sharecare.com/static/realage
Rejuvant; www.rejuvant.com
Replika; www.replika.ai
resTORbio; www.restorbio.com
Reversible Destiny Foundation; www.reversibledestiny.org
Roam Robotics; www.roamrobotics.com
Samumed; www.samumed.com
SIWA Therapeutics; www.siwatherapeutics.com
Spacetime Ventures; www.spacetimeventures.com
SpectraCell Laboratories; www.spectracell.com

Synthego; www.synthego.com
Teijin's The Next 100 Think Human Exhibition; 100.teijin.co.jp
Terasem Movement Foundation; www.terasemmovementfoundation.com
Timeship; www.timeship.org
Transhumanist Party; www.transhumanist-party.org/
Transhumanity.net; www.transhumanity.net
Unity Biotechnology; www.unitybiotechnology.com
Viome; www.viome.com
Virtual Human Interaction Lab, Stanford University; www.vhil.stanford.edu

Photography Credits

Several photographs throughout this book were sourced from Dreamstime.com, available under Creative Commons Zero (CC0) License. Thank you to the photographers and individuals who support open source photography organizations.

Acknowledgments

Alice in Futureland would not have been possible without the guidance of numerous individuals who in one way or another have contributed their valuable innovative wisdom and energy.

First and foremost, we owe our deepest gratitude to our Sputnik Futures colleagues and partners, Lisa, Amy, and Jordan, who have wandered with us through the many years of futures research work. Know that we have circled the stars together, and have many more laps to go. ☺

Thank you, Luis, for bringing Alice to life.

Special thanks to the like minds at Tiller Press, especially Theresa, who saw the vision for this future series, Sam for his insights, and Emily for her sharp editing.

And to the many frontier thinkers who have shared their knowledge with enthusiasm, we are gratefully indebted.

To our families that have lived this journey with us, thank you all with love!

–J&J

Endnotes

1. "Clarke's Three Laws," *World Heritage Encyclopedia*, http://self.gutenberg.org/articles/eng/Clarke%27s_three_laws.

01: THERE'S A NEW DISEASE CALLED "AGING"; DO YOU HAVE IT?

1. Elysium Health press release; Centers for Disease Control and Prevention, 2013, https://www.cdc.gov/aging/pdf/state-aging-health-in-america-2013.pdf.
2. Alvin Powell, "Longevity and Anti-aging Research: 'Prime Time for an Impact on the Globe,'" *Harvard Gazette*, March 8, 2019, https://news.harvard.edu/gazette/story/2019/03/anti-aging-research-prime-time-for-an-impact-on-the-globe/.
3. Michael Rose, "Evolutionary Biology of Diet, Aging and Mismatch," Ancestry Foundation, 2018.
4. Anthony King, "Can We Live Forever?" Chemistry World, January 28, 2019.
5. Blanka Rogina, Robert A. Reenan, Steven P. Nilsen, and Stephen L. Helfand, "Extended Life-span Conferred by Cotransporter Gene Mutations in Drosophila," *Science* 290, no. 5499 (December 15, 2000): 2137–40, https://science.sciencemag.org/content/290/5499/2137.full.
6. Nir Barzilai, Jill P. Crandall, Stephen B. Kritchevsky, and Mark A. Espeland, "Metformin as a Tool to Target Aging," *Cell Metabolism* 23, no. 6 (June 14, 2016): 1060–65, doi: 10.1016/j.cmet.2016.05.011.
7. Cell Press, "How Exercise—Interval Training in Particular—Helps Your Mitochondria Stave Off Old Age," *ScienceDaily*, March 7, 2017, https://www.sciencedaily.com/releases/2017/03/170307155214.htm.
8. John Fawkes, "Is Autophagy the Secret to Life Extension?" *Medium*, August 5 (n.d.), https://medium.com/better-humans/is-autophagy-the-secret-to-life-extension-cbaf4fe54be2.

9. B. G. Childs, M. Gluscevic, D. J. Baker, et al., "Senescent Cells: An Emerging Target for Diseases of Ageing," *Nature Reviews Drug Discovery* 16, no. 10 (October 2017): 718–35, doi: 10.1038/nrd.2017.116.
10. M. Xu, T. Pirtskhalava, J. N. Farr, et al., "Senolytics Improve Physical Function and Increase Life Span in Old Age," *Nature Medicine* 24, no. 8 (August 2018): 1246–56, doi: 10.1038/s41591-018-0092-9.
11. Ibid., cited on Medical Xpress, July 9, 2018.
12. Jamie N. Justice, Anoop M. Nambiar, Tamar Tchkonia, Nathan K. LeBrasseur, Rodolfo Pascual, Shahrukh K. Hashmi, Larissa Prata, Michal M. Masternak, Stephen B. Kritchevsky, Nicolas Musi, and James L. Kirkland, "Senolytics in Idiopathic Pulmonary Fibrosis: Results from a First-in-Human, Open-Label Pilot Study," *EBioMedicine* 40 (February 2019): 554–63, doi: https://doi.org/10.1016/j.ebiom.2018.12.052.
13. King, "Can We Live Forever?"
14. Mayo Clinic Staff, "Stem Cells: What They Are and What They Do," MayoClinic.org, June 8, 2019.
15. "The Convenient Truth About Stem Cells and Anti-Aging," RAADfest, July 10, 20019, https://www.raadfest.com/raad-fest/convenient-truth-stem-cells-and-anti-aging.
16. Powell, "Longevity and Anti-aging Research."
17. Reason, "Senolytic Treatment with Dasatinib and Quercetin Confirmed to Reduce the Burden of Senescent Cells in Human Patients," FightAging.org, September 18, 2019, https://www.fightaging.org/archives/2019/09/senolytic-treatment-with-dasatinib-and-quercetin-confirmed-to-reduce-the-burden-of-senescent-cells-in-human-patients/.
18. "Age Reversal Takeaways from RAADfest 2018: What We Can Do Right Now to Increase Vitality and Reduce Aging Biomarkers," People Unlimited, https://peopleunlimitedinc.com/posts/2018/11/19/age-reversal-takeways-from-raadfest-2018.

02: WHAT'S YOUR GRIMAGE?

1. Yannick Stephan, Angelina R. Sutin, and Antonio Terracciano, "Subjective Age and Mortality in Three Longitudinal Samples," *Psychoso-*

matic Medicine 80, no. 7 (September 2018): 659–64, doi: 10.1097/PSY.0000000000000613.

2. Annibale Alessandro Puca, Albino Carrizzo, Chiara Spinelli, et al., "Single Systemic Transfer of a Human Gene Associated with Exceptional Longevity Halts the Progression of Atherosclerosis and Inflammation in ApoE Knockout Mice through a CXCR4-Mediated Mechanism," *European Heart Journal* (July 10, 2019), ehz459, https://doi.org/10.1093/eurheartj/ehz459.
3. Terra Marquette, "'Longevity Gene' Protein Offers Possibility of Extended, Healthier Lives," Study Finds, July 25, 2019, https://www.studyfinds.org/longevity-gene-protein-offers-possibility-extended-healthier-lives/.
4. Umeå University, "Depression and Chronic Stress Accelerates Aging," *ScienceDaily*, November 10, 2011, https://www.sciencedaily.com/releases/2011/11/111109093729.htm.
5. Arizona State University, "Hidden Secret of Immortality Enzyme Telomerase: Can We Stay Young Forever, or Even Recapture Lost Youth?" *ScienceDaily*, February 27, 2018, https://www.sciencedaily.com/releases/2018/02/180227142114.htm.
6. Yinnan Chen, Joshua D. Podlevsky, Dhenugen Logeswaran, and Julian J.-L. Chen, "A Single Nucleotide Incorporation Step Limits Human Telomerase Repeat Addition Activity," *Embo Journal* 37, no. 6 (March 15, 2018): e97953, https://www.embopress.org/doi/pdf/10.15252/embj.201797953.
7. Mind and Life Institute, "The Telomere Effect: A Revolutionary Approach to Living Younger, Healthier, Longer," MindandLife.org, https://www.mindandlife.org/books/telomere-effect-revolutionary-approach-living-younger-healthier-longer/.
8. John Zarocostas, "Chronic Diseases among the Over 40s in China Are Set to Double over the Next 20 Years," *BMJ* (July 27, 2011): 343, https://doi.org/10.1136/bmj.d4801.
9. Christine Buttorff, Teague Ruder, and Melissa Bauman, "Multiple Chronic Conditions in the United States," RAND Corporation, 2017, https://doi.org/10.7249/TL221.
10. Dongjuan Dai, Aaron J. Prussin II, Linsey C. Marr, Peter J. Vikesland,

Marc A. Edwards, and Amy Pruden, "Factors Shaping the Human Exposome in the Built Environment: Opportunities for Engineering Control," *Environmental Science & Technology* 51, no 14 (2017): 7759–74, https://pubs.acs.org/doi/full/10.1021/acs.est.7b01097.

11. Eni Gómez, Eva Toribio, Montserrat Delor, and Judit Morlà Folch, "Hair Exposome, a Concept to Address Hair Disorders from a New Perspective by Beauty Cluster Barcelona," Skinobs Cosmetic Testing News, February 28, 2018, https://skinobs.com/news/en/non-classe en/hair-exposome-a-concept-to-address-hair-disorders-from-a-new -perspective-by-beauty-cluster/.
12. Syam S. Andra, Christine Austin, Robert O. Wright, and Manish Arora, "Reconstructing Pre-natal and Early Childhood Exposure to Multi-Class Organic Chemicals Using Teeth: Towards a Retrospective Temporal Exposome," *Environment International* 83 (October 2015): 137–45, https://www.ncbi.nlm.nih.gov/pmc/articles/PMC4545311/.
13. Shi Ying, Dan-Ning Zeng, Liang Chi, Yuan Tan, Carlos Galzote, Cesar Cardona, Simon Lax, Jack Gilbert, and Zhe-Xue Quan, "The Influence of Age and Gender on Skin-Associated Microbial Communities in Urban and Rural Human Populations," *PLoS ONE* (October 28, 2015), https://journals.plos.org/plosone/article?id=10.1371/journal .pone.0141842.
14. Colleen Zacharyczuk, "Human Microbiome Project May Hold Promise for the Future," Healio, June 2012, https://www.healio.com/pediatrics /news/print/infectious-diseases-in-children/%7B39d199b9-64c6 -4345-90fa-f96293fb258a%7D/human-microbiome-project-may -hold-promise-for-future.
15. Brian Parker, "Circadian Rhythm, Sleep, and Aging," *Life Extension Magazine*, 2018, https://www.lifeextension.com/magazine/2018/ss /circadian-rhythm-sleep-and-aging.
16. Beatrice Bretherton, Lucy Atkinson, Aaron Murray, Jennifer Clancy, Susan Deuchars, and Jim Deuchars, "Effects of Transcutaneous Vagus Nerve Stimulation in Individuals Aged 55 Years or Above: Potential Benefits of Daily Stimulation," *Aging* 11, no. 14 (July 31, 2019): 4836–57, https://doi.org/10.18632/aging.102074.
17. S. Horvath, "DNA Methylation Age of Human Tissues and Cell Types,"

Genome Biology 14, no. 10 (2013): R115, doi: 10.1186/gb-2013-14-10-r115.
18. S. Horvath, J. Oshima, G. M. Martin, et al., "Epigenetic Clock for Skin and Blood Cells Applied to Hutchinson Gilford Progeria Syndrome and Ex Vivo Studies," *Aging (Albany, NY)* 10, no. 7 (July 26, 2018): 1758–75, doi: 10.18632/aging.101508.
19. Joris Deelen, Johannes Kettunen, and P. Eline Slagboom, "A Metabolic Profile of All-Cause Mortality Risk Identified in an Observational Study of 44,168 Individuals," *Nature Communications* 10, Article number 3346 (2019), https://www.nature.com/articles/s41467-019-11311-9.
20. Helen Thomson, "Drug Cocktail Seems to Reverse Biological Signs of Ageing in People," *New Scientist*, September 5, 2019, https://www.newscientist.com/article/2215537-drug-cocktail-seems-to-reverse-biological-signs-of-ageing-in-people/.
21. Gregory M. Fahy, Robert T. Brooke, James P. Watson, et al., "Reversal of Epigenetic Aging and Immunosenescent Trends in Humans," *Aging Cell* (September 8, 2019), https://doi.org/10.1111/acel.13028.
22. Kate Whitehead, "How to Look Younger Than You Are: Biomarkers—Which to Track and How to Step In if Needed," *South China Morning Post*, August 5, 2019, https://www.scmp.com/lifestyle/health-wellness/article/3021225/how-look-younger-you-are-biomarkers-which-track-and-how.
23. Lauren Bramley, MD, "Know Your Numbers," TED Talk, TEDXVail, January 25, 2016.
24. Whitehead, "How to Look Younger Than You Are."

03: IS NATURE IMMORTAL?

1. David Brin, PhD, "Do We Really Want Immortality?" *iPlanet*, 1999.
2. Stephen S. Hall, "Finally, the Drug That Keeps You Young," *MIT Technology Review*, October 23, 2018.
3. Yang Zhao, Alexander Tyshkovskiy, Daniel Muñoz-Espín, Xiao Tian, Manuel Serrano, Joao Pedro de Magalhaes, Eviatar Nevo, Vadim N. Gladyshev, Andrei Seluanov, and Vera Gorbunova, "Cellular Senescence in the Naked Mole Rat," *Proceedings of the National Academy of Sciences of the United States of America* 115, no. 8 (February

2018): 1801–6, https://doi.org/10.1073/pnas.1721160115.

4. Geoffrey Mohan, "Naked Mole Rat May Be Ugly, but It Could Hold Secret to Longevity," *Los Angeles Times*, October 1, 2013, https://www.latimes.com/science/sciencenow/la-sci-sn-mole-rat-longevity-20131001-story.html.
5. University of Nottingham, "Immortal Worms Defy Aging," *ScienceDaily*, February 27, 2012, www.sciencedaily.com/releases/2012/02/120227152612.htm.
6. Luca Gentile, Francesc Cebrià, and Kerstin Bartscherer, "The Planarian Flatworm: An In Vivo Model for Stem Cell Biology and Nervous System Regeneration," *Disease Models & Mechanisms* 4 (2011): 12–19, https://dmm.biologists.org/content/4/1/12; and Kerstin Bartscherer, "Flatworms, the Masters of Regeneration—but Nothing Can Happen Without Stem Cells," Max Planck Institute for Molecular Biomedicine (June 3, 2014), https://www.mpg.de/8244494/flatworms-regeneration##targetText=Flatworms%2C%20the%20masters%20of%20regeneration%20%E2%80%93%20but%20nothing,can%20happen%20without%20stem%20cells&targetText=Planarians%20are%20known%20as%20masters,number%20of%20pluripotent%20stem%20cells.
7. Anna-Marei Boehm, Konstantin Khalturin, Friederike Anton-Erxleben, et al., "FoxO Is a Critical Stem Cell Regulator in Hydra," *Proceedings of the National Academy of Sciences of the United States of America* 109, no. 48 (November 2012): 19697–702, https://doi.org/10.1073/pnas.1209714109.
8. S. D. Tardif, K. G. Mansfield, R. Ratnam, C. N. Ross, and T. E. Ziegler, "The Marmoset as a Model of Aging and Age-Related Diseases," *ILAR Journal* 52, no. 1 (2011): 54–65, doi: 10.1093/ilar.52.1.54.
9. Judith A. Mattison, Ricki J. Colman, T. Mark Beasley, et al., "Caloric Restriction Improves Health and Survival of Rhesus Monkeys," *Nature Communications* 8, article number 14063 (2017), https://www.nature.com/articles/ncomms14063.
10. L. M. Redman and E. Ravussin, "Caloric Restriction in Humans: Impact on Physiological, Psychological, and Behavioral Outcomes," *Antioxidants & Redox Signaling* 14, no. 2 (January 15, 2011): 275-87, doi: 10.1089/ars.2010.3253.

11. Matthew Blakeslee, "Long-Lived Yeast May Hold a Key to Human Longevity," *USC News*, April 20, 2001, https://news.usc.edu/5375/Long-lived-yeast-may-hold-a-key-to-human-longevity/.
12. Matthew J. Yousefzadeh, Yi Zhu, Sara J. McGowan, et al., "Fisetin Is a Senotherapeutic That Extends Health and Life Span," *EBioMedicine* 36 (October 2018): 18–28, https://www.ncbi.nlm.nih.gov/pmc/articles/PMC6197652/.
13. Yuki Yoshida, Georgios Koutsovoulos, Dominik R. Laetsch, et al., "Comparative Genomics of the Tardigrades *Hypsibius dujardini* and *Ramazzottius varieornatus*," *PLoS Biology* 15, no. 7 (July 27, 2017), https://doi.org/10.1371/journal.pbio.2002266.
14. Ibid.
15. National Academies of Sciences, Engineering, and Medicine, *Biodefense in the Age of Synthetic Biology* (Washington, DC: The National Academies Press, 2018).
16. Kendall K. Morgan, PhD, "Synthetic Biology Goes to Nature . . . and Beyond," *Genetic Engineering & Biotechnology News*, August 1, 2019.
17. Salk Institute, "CRISPR/Cas9 Therapy Can Suppress Aging, Enhance Health and Extend Life Span in Mice," *ScienceDaily*, February 19, 2019, https://www.sciencedaily.com/releases/2019/02/190219111747.htm##targetText=CRISPR%2FCas9%20therapy%20can%20suppress%20aging%2C%20enhance%20health%20and,extend%20life%20span%20in%20mice&targetText=Summary%3A,help%20decelerate%20the%20aging%20process.&targetText=Now%2C%20Salk%20Institute%20researchers%20have,help%20decelerate%20the%20aging%20process.
18. Alexis C. Komor, Yongjoo B. Kim, Michael S. Packer, John A. Zuris, and David R. Liu, "Programmable Editing of a Target Base in Genomic DNA Without Double-Stranded DNA Cleavage," *Nature* 533, no. 7603 (May 19, 2016): 420–24, https://www.nature.com/articles/nature17946.
19. John Rennie and Lucy Reading-Ikkanda, "Seeing the Beautiful Intelligence of Microbes," *Quanta Magazine*, November 13, 2017, https://www.quantamagazine.org/the-beautiful-intelligence-of-bacteria-and-other-microbes-20171113/.
20. Steven T. Bruckbauer, Joseph D. Trimarco, Joel Martin, et al., "Ex-

perimental Evolution of Extreme Resistance to Ionizing Radiation in Escherichia coli after 50 Cycles of Selection," *Journal of Bacteriology* 201, no. 8 (March 2019), https://jb.asm.org/content/201/8/e00784-18.

21. Freeman Dyson, "Our Biotech Future," *New York Review of Books*, July 19, 2007.

04: WILL ONE HUNDRED BE THE NEW FORTY?

1. Erika Hayasaki, "The Mouse That Died of Young Age," *MIT Technology Review* 122, no. 5 (September-October 2019): 24–29.
2. Max Roser, "Life Expectancy," OurWorldInData.org, 2019, https://our worldindata.org/life-expectancy.
3. "The Baby Bonanza," *Economist*, August 27, 2009, https://www.econo mist.com/briefing/2009/08/27/the-baby-bonanza.
4. Jane Wollman Rusoff, "Moshe Milevsky: Are You as 'Old' as You Think You Are?" ThinkAdvisor, March 22, 2019.
5. Lewis Braham, "Forget Your Real Age, Plan Your Retirement around Your 'Biological Age,'" *Barron's*, January 27, 2019.
6. Christopher Wareham, "How Can Life-Extending Treatments Be Available for All?" Aeon, August 2, 2017, https://aeon.co/ideas/how-can -life-extending-treatments-be-available-for-all.
7. "New England Centenarian Study," Boston University School of Medicine, https://www.bumc.bu.edu/centenarian/.
8. Peter Martin, Grace da Rosa, Ilene C. Siegler, Adam Davey, Maurice MacDonald, and Leonard W. Poon, "Personality and Longevity: Findings from the Georgia Centenarian Study," *Age* 28, no. 4 (December 2006): 343–52; PMC, November 21, 2006, https://www.ncbi.nlm.nih .gov/pmc/articles/PMC3259159/##targetText=Results%20of%20 a%20personality%20study,younger%20groups%20(Martin%202002).
9. New England Centenarian Study, "Why Study Centenarians? An Overview," Boston University School of Medicine, https://www.bumc.bu .edu/centenarian/overview/.
10. Dave Asprey, "This Entrepreneur Plans to Live to 180—Here Are His 5 Health Hacks," *Entrepreneur*, December 20, 2017.
11. Michel Poulain, Giovanni Pes, Claude Grasland, et al., "Identification of a Geographic Area Characterized by Extreme Longevity in the Sardinia Islands: The AKEA study," *Experimental Gerontology* 39,

no. 9 (2004): 1423–29, https://halshs.archives-ouvertes.fr/halshs-00175541/document.
12. D. Craig Willcox, Bradley J. Willcox, Qimei He, Nien-chiang Wang, and Makoto Suzuki, "They Really Are That Old: A Validation Study of Centenarian Prevalence in Okinawa," *Journals of Gerontology: Series A* 63, no. 4 (April 2008): 338–49, https://doi.org/10.1093/gerona/63.4.338.
13. Dante A. Ciampaglia, "These Architects Sought to Solve the Ultimate Human Design Flaw—Death," *Metropolis*, May 30, 2018.
14. Reversible Destiny Foundation, http://www.reversibledestiny.org.
15. Eliezer Yudkowsky, "Fun Theory," Less Wrong Wiki, June 2017, https://wiki.lesswrong.com/wiki/Fun_theory.

05: WILL WE BECOME SOFTWARE?

1. Alastair Himmer, "Kyoto Temple Puts Faith in Robot Priest," *Japan Times*, August 15, 2019, https://www.japantimes.co.jp/news/2019/08/15/business/tech/kyoto-temple-robot-priest/.
2. Martine Rothblatt, "Mindclones in 4 Easy Pieces," *Mindfiles, Mindware, and Mindclones* (blog), October 4, 2013, http://www.mindclones.blogspot.com.
3. Carl J. Öhman and David Watson, "Are the Dead Taking Over Facebook? A Big Data Approach to the Future of Death Online," *Big Data & Society* (April 23, 2019), https://doi.org/10.1177/2053951719842540.
4. "Making It Easier to Honor a Loved One on Facebook after They Pass Away," Facebook.com, April 9, 2019.
5. Öhman and Watson, "Are the Dead Taking Over Facebook?"
6. George Orwell, *1984* (ebook edition: https://www.planetebook.com/free-ebooks/1984.pdf), 313–14.
7. Carl J. Öhman and Luciano Floridi, "An Ethical Framework for the Digital Afterlife Industry," *Nature Human Behavior* 2 (2018): 318–20, https://www.nature.com/articles/s41562-018-0335-2.
8. Martine Rothblatt, PhD, *Virtually Human: The Promise & Peril of Digital Immortality* (New York: Picador, 2015).
9. Clifford N. Lazarus, PhD, "Does Consciousness Exist Outside of the

Brain?" *Psychology Today*, June 26, 2019, https://www.psychology today.com/us/blog/think-well/201906/does-consciousness-exist -outside-the-brain.

10. Ibid.; Peter Fenwick and Elizabeth Fenwick, *The Art of Dying* (London: Bloomsbury, 2008).
11. Sam Parnia, Ken Spearpoint, Gabriele de Vos, et al., "AWARE—AWAreness during REsuscitation—A Prospective Study," *Resuscitation* 85, no. 13 (December 2014), doi: 10.1016/j.resuscitation.2014.09.004.
12. Alex Johnson, "Elon Musk Wants to Hook Your Brain Directly Up to Computers—Starting Next Year," NBCnews.com, July 17, 2019.
13. "iHuman Perspective: Neural Interfaces," Royal Society (September 10, 2019), https://royalsociety.org/topics-policy/projects/ihuman -perspective/.
14. Ian Sample, "'Neural Revolution': Royal Society Calls for Inquiry into New Wave of Brain Implants," *Guardian*, September 10, 2019, https:// www.theguardian.com/science/2019/sep/10/neural-revolution -royal-society-calls-for-inquiry-into-new-wave-of-brain-implants## targetText="People%20could%20become%20telepathic%20to,friends %2C"%20the%20report%20states.
15. Eric Mack, "Former Facebook Exec Wants to Make Us Telepathic by 2025," Cnet.com, July 9, 2017.
16. Robbin A. Miranda,William D. Casebeer, Amy M. Hein, et al., "DARPA-Funded Efforts in the Development of Novel Brain-Computer Interface Technologies," *Journal of Neuroscience Methods* 244 (April 5, 2015): 52–67, https://www.sciencedirect.com/science/article/pii /S0165027014002702?via%3Dihub.
17. R. Peters, "Ageing and the Brain," *Postgraduate Medical Journal* 82, no. 964 (February 2006): 84–88, doi: 10.1136/pgmj.2005.036665.
18. http://www.biocybernaut.com.
19. Ramez Naam, "How to Turn Science Fiction into Science Fact," Neo .Life, updated October 15, 2019.
20. Connie X. Wang, Isaac A. Hilburn, Daw-An Wu, et al., "Transduction of the Geomagnetic Field as Evidenced from Alpha-Band Activity in the Human Brain," *eNeuro* 6, no. 2 (March 18, 2019), doi: https://doi .org/10.1523/ENEURO.0483-18.2019.

06: WHAT WILL IT MEAN TO BE "HUMAN"?

1. Max More, "Principles of Extropy (Version 3.11): An Evolving Framework of Values and Standards for Continuously Improving the Human Condition," Extropy Institute, archived from the original on October 15, 2013.
2. Christianna Reedy, "Kurzweil Claims the Singularity Will Happen by 2045: Get Ready for Humanity 2.0," *Futurism*, October 5, 2017.
3. Jo Marchant, "Immunology: The Pursuit of Happiness," *Nature News*, November 27, 2013.
4. Jennifer E. Graham, Lisa M. Christian, and Janice K. Kiecolt-Glaser, "Stress, Age, and Immune Function: Toward a Lifespan Approach," *Journal of Behavioral Medicine* 29, no. 4 (August 2006): 389–400; PMC, May 19, 2006, https://www.ncbi.nlm.nih.gov/pmc/articles/PMC2805089/.
5. Lewina O. Lee, Peter James, Emily S. Zevon, et al., "Optimism Is Associated with Exceptional Longevity in 2 Epidemiologic Cohorts of Men and Women," *Proceedings of the National Academy of Sciences of the United States of America* 116, no. 37 (September 2019): 18357–62, https://doi.org/10.1073/pnas.1900712116.
6. Becca R. Levy, Martin D. Slade, Robert H. Pietrzak, and Luigi Ferrucci, "Positive Age Beliefs Protect Against Dementia Even Among Elders with High-Risk Gene," *PLoS ONE*, February 7, 2018, https://journals.plos.org/plosone/article?id=10.1371/journal.pone.0191004.
7. J. H. Fowler and N. A. Christakis, "Dynamic Spread of Happiness in a Large Social Network: Longitudinal Analysis over 20 Years in the Framington Heart Study," *British Medical Journal* 337 (December 5, 2008): a2338.
8. The Framingham Heart Study is a project of Boston University and the National Heart, Lung, and Blood Institute.
9. Anatoli I. Yashin, Igor V. Akushevich, Konstantin G. Arbeev, Lucy Akushevich, Svetlana V. Ukraintseva, and Aliaksandr Kulminski, "Insights on Aging and Exceptional Longevity from Longitudinal Data: Novel Findings from the Framingham Heart Study," *Age* 28, no. 4 (2006): 363–74, doi: 10.1007/s11357-006-9023-7.
10. Nicholas A. Christakis, "How AI Will Rewire Us," *Atlantic*, April 2019.

11. Zoltan Istvan, "The Three Laws of Transhumanism and Artificial Intelligence," *Psychology Today*, September 29, 2014.
12. The Next 100 Think Human Project, "Humanity Lies in Being Able to Change Our Bodies and Its Roles," Think 01, Vol. 3, Teijin, https://100.teijin.co.jp/en/project01/vol3/.
13. Dr. Al Emondi, "Next-Generation Nonsurgical Neurotechnology," DARPA, https://www.darpa.mil/program/next-generation-nonsurgical-neurotechnology.
14. Julian Huxley, *Religion Without Revelation* (London: E. Benn, 1927). Quoted in James Hughes, *Citizen Cyborg: Why Democratic Societies Must Respond to the Redesigned Human of the Future* (Cambridge, MA: Westview Press, 2004).
15. Nick Bostrom, "A History of Transhumanist Thought," originally published in *Journal of Evolution and Technology* 14, no. 1 (April 2005), reprinted in slightly edited form at https://www.nickbostrom.com/papers/history.pdf.
16. Theødor, "Natasha Vita-More Doesn't Want to Play God. She Just Wants to Play," *Medium*, September 2, 2017.
17. Ames Research Center, NASA.
18. Michio Kaku, "How Your Immortal Consciousness Will Travel the Universe," video and transcript, Big Think, July 29, 2019, https://bigthink.com/videos/michio-kaku-2639431116.

07: ALICE IN FUTURELAND

1. Jaron Lanier, *You Are Not a Gadget: A Manifesto* (New York: Vintage, 2011).
2. Jean Huston, *The Possible Human: A Course in Enhancing Your Physical, Mental, and Creative Abilities* (New York: TarcherPerigee, 1997).
3. Jay W. Richards, ed., *Are We Spiritual Machines? Ray Kurzweil vs. the Critics of Strong A.I.* (Discovery Institute, 2001).

Illustration Credits

Several photographs throughout this book were sourced from Dreams time.com, available under the Creative Commons Zero (CC0) license. Thank you to the photographers and individuals who support open-source photography organizations.

INTRODUCTION

pp. viii–ix (*clockwise from upper left*): Digital Pill: AdobeStock_70178127; Grim Reaper: AdobeStock_92845141; Ringed Worm: Free photo 95887469 © Publicdomainphotos–Dreamstime.com, https://www.dreamstime.com/terrestrial-animal-close-up-macro-photography-ringed-worm-public-domain-image-free-95887469; Two Cells: AdobeStock_203494644; Grid over Face: AdobeStock_168579073; Wooden Model with Pencils: Free photo 4593758 © Michael Flippo–Dreamstime.com, https://www.dreamstime.com/drawing-tools-free-stock-photos-image-free-4593758

p. xi: Ostrich: Free photo 82985170 © creativecommonsstockphotos–Dreamstime.com, https://www.dreamstime.com/ostrich-portrait-against-sky-public-domain-image-free-82985170

CHAPTER 01: THERE'S A NEW DISEASE CALLED "AGING"; DO YOU HAVE IT?

p. xii: Digital Pill: AdobeStock_70178127

p. 3: Pill Container: AdobeStock_246446506

pp. 4, 7: Red and White Chairs: AdobeStock_159064656

pp. 8–9: Two Arms: AdobeStock_235283416

pp. 10, 13: Empty Dinner Plate: AdobeStock_272726606

p. 14: Fruit Fly: AdobeStock_196689445

p. 17: Pills: Free photo 128440423 © Publicdomainphotos–Dreamstime.com, https://www.dreamstime.com/product-drug-font-pill-public-domain-image-free-128440423

p. 18: Moai: Free photo 92129182 © publicdomainstockphotos–Dreamstime.com, https://www.dreamstime.com/moia-added-sky-small-pottery-head-inches-high-my-easter-island-thing-deepest-leicester-city-uk-s-just-bit-public-domain-image-free-92129182

pp. 20–21: Cell: AdobeStock_232975938
p. 22: Man Exercising: AdobeStock_166747861
pp. 25, 26: Washing Machine: AdobeStock_101620847
pp. 28, 31: Room with Curtains: Free photo 109915472 © creativecommonsstockphotos–Dreamstime.com, https://www.dreamstime.com/architecture-clean-curtains-public-domain-image-free-109915472
p. 32: Man Looking in Mirror: Free photo 83010280 © creativecommonsstockphotos–Dreamstime.com, https://www.dreamstime.com/blue-eyed-man-staring-mirror-public-domain-image-free-83010280
p. 35: Clouds Reflected in Mirror: Free photo 82994562 © creativecommonsstockphotos–Dreamstime.com, https://www.dreamstime.com/clouds-sky-reflected-mirror-round-lying-grass-reflecting-blue-public-domain-image-free-82994562
p. 36: Hand with Two Cells: AdobeStock_249650231
pp. 38–39: Electric Sockets and Plugs: Free photo 83008731 © creativecommonsstockphotos–Dreamstime.com, https://www.dreamstime.com/black-socket-white-switch-besides-converter-public-domain-image-free-83008731

CHAPTER 02: WHAT'S YOUR GRIMAGE?
p. 40: Grim Reaper: AdobeStock_92845141
pp. 42–43: Pocket Watch: Free photo 82978395 © creativecommonsstockphotos–Dreamstime.com, https://www.dreamstime.com/hanging-watch-old-fashioned-pale-blue-background-public-domain-image-free-82978395
pp. 44–45: Matchsticks: AdobeStock_251213505
p. 47: Decor: yellow-and-pink-lighted-x-decor-1964471
p. 49: *The Telomere Effect* Book Cover: 99110866249cc86f8e07f5d100b82e0be14f8614-book
p. 50: Yellow Symbol: local.png
p. 52: Internal Organs: AdobeStock_208980048
p. 55: Man with Earphones: Free photo 91252687 © creativecommonsstockphotos–Dreamstime.com, https://www.dreamstime.com/man-subway-putting-earphones-ears-back-portrait-young-his-public-domain-image-free-91252687

pp. 56–57: Wall Clock: Free photo 106174011 © creativecommons stockphotos–Dreamstime.com, https://www.dreamstime.com/wall-clock -public-domain-image-free-106174011
p. 58: Person Wearing Watch: Free photo 115628423 © creative commonsstockphotos–Dreamstime.com, https://www.dreamstime.com /person-wearing-round-analog-watch-public-domain-image-free-115 628423
pp. 63, 64: Fingerprints: Free photo 400963 © Specular–Dreamstime com, https://www.dreamstime.com/bloody-fingerprints-stock-photos -image-free-400963
pp. 67, 68: Bicycle Chain: Free photo 116504709 © creativecommons stockphotos–Dreamstime.com, https://www.dreamstime.com/gray -bicycle-chain-orange-surface-public-domain-image-free-116504709

CHAPTER 03: IS NATURE IMMORTAL?

p. 70: Ringed Worm: Free photo 95887469 © Publicdomainphotos–Dreamstime.com, https://www.dreamstime.com/terrestrial-animal-close -up-macro-photography-ringed-worm-public-domain-image-free-958 87469
p. 73: Pink Jellyfish: Free photo 114321102 © creativecommonsstock photos–Dreamstime.com, https://www.dreamstime.com/purple-pink -jelly-fish-public-domain-image-free-114321102
p. 74: Seashells: Free photo 119034936 © Publicdomainphotos–Dreamstime.com, https://www.dreamstime.com/clam-cockle-seashell -clams-oysters-mussels-scallops-public-domain-image-free-119034936
p. 76: Power Symbol: 184-1846880_banner-library-download-computer -icons-portable-network-power
p. 77: Ontogeny Diagram: Ontogeny-reversal-in-the-hydromedusa -Turritopsis-nutricula-After-S-Piraino-et-al
p. 78: Close-Up of Tortoise Foot: Free photo 100631142 © Publicdomain photos–Dreamstime.com, https://www.dreamstime.com/terrestrial -animal-mammal-fauna-close-up-public-domain-image-free-100631142
p. 82: Monkey Portrait: Free photo 96793620 © creativecommonsstock photos–Dreamstime.com, https://www.dreamstime.com/portrait-monkey -public-domain-image-free-96793620

p. 84: Cells: AdobeStock_21047189
pp. 86–87: Tardigrade: iStock-870735250
p. 89: Digital Petri Dish: AdobeStock_201629830
p. 91: Highlighted Eye: Free photo 100644865 © Publicdomainphotos–Dreamstime.com, https://www.dreamstime.com/face-eyebrow-skin-eye-public-domain-image-free-100644865
pp. 92–93: Microbes: Free photo 7031914 © Dmitry Sunagatov–Dreamstime.com, https://www.dreamstime.com/microbe-stock-images-image-free-7031914
p. 95: Woman in Hoodie: Free photo 119611327 © creativecommonsstockphotos–Dreamstime.com, https://www.dreamstime.com/woman-wearing-hooded-pullover-hoodie-facing-tablet-computer-public-domain-image-free-119611327
pp. 96–97: Round Mirrors: Free photo 109911754 © creativecommonsstockphotos–Dreamstime.com, https://www.dreamstime.com/two-round-mirror-beige-frames-public-domain-image-free-109911754

CHAPTER 04: WILL ONE HUNDRED BE THE NEW FORTY?

p. 98: Wooden Model with Pencils: Free photo 4593758 © Michael Flippo–Dreamstime.com, https://www.dreamstime.com/drawing-tools-free-stock-photos-image-free-4593758
p. 100: Birthday Candles: Free photo 116050142 © creativecommonsstockphotos–Dreamstime.com, https://www.dreamstime.com/red-green-birthday-candle-lights-public-domain-image-free-116050142
p. 101: New Jerusalem Missionary Baptist Church Sign: AdobeStock_154087746_Editorial_Use_Only
p. 103: Baby: Free photo 109918078 © creativecommonsstockphotos–Dreamstime.com, https://www.dreamstime.com/close-up-photography-baby-looking-public-domain-image-free-109918078
p. 104: Three Women: Pixabay: All content on Pixabay can be used for free for commercial and noncommercial use across print and digital; https://pixabay.com/es/photos/familia-mujeres-blanco-y-negro-2492164/
pp. 106–7: Man with Red Umbrella: AdobeStock_169411306
p. 109: Woman with Red Umbrella: Free photo 115628390 © creativecommonsstockphotos–Dreamstime.com, https://www.dreamstime.com

/woman-wearing-black-long-sleeved-dress-holding-red-umbrella-public-domain-image-free-115628390
p. 110: Man Looking in Mirror: Free photo 118290540 © creative commonsstockphotos–Dreamstime.com, https://www.dreamstime.com/photo-man-looking-mirror-public-domain-image-free-118290540
p. 113: Yellow Flower with Mirror: Free photo 115013079 © creative commonsstockphotos–Dreamstime.com, https://www.dreamstime.com/yellow-petaled-flower-mirror-public-domain-image-free-115013079
p.114: Couple Holding Hands: Free photo 15223415 © Melinda Nagy–Dreamstime.com, https://www.dreamstime.com/senior-couple-s-hands-free-stock-photo-image-free-15223415
p. 115: Exercising Seniors: Free photo 83000613 © publicdomain stockphotos–Dreamstime.com, https://www.dreamstime.com/exercising-seniors-group-senior-citizens-resistance-bands-chairs-public-domain-image-free-83000613
p. 116: Red Mug: Free photo 112301402 © creativecommonsstock photos–Dreamstime.com, https://www.dreamstime.com/red-mug-string-lights-public-domain-image-free-112301402
p. 117: Falling Pills: Free photo 90081752 © Publicdomainphotos–Dreamstime.com, https://www.dreamstime.com/orange-red-text-computer-wallpaper-public-domain-image-free-90081752
p. 118: Cobblestones: AdobeStock_74665936
p. 119: People Swimming: Free photo 82959788 © creativecommons stockphotos–Dreamstime.com, https://www.dreamstime.com/swimmers-blue-coastal-waters-malta-sunny-day-public-domain-image-free-82959788
p. 120: Salad Plate: AdobeStock_191240027
p. 122: Pinwheels: Free photo 119114456 © creativecommonsstock photos–Dreamstime.com, https://www.dreamstime.com/selective-focus-photography-pinwheels-string-public-domain-image-free-119114456
p. 124: Boy Climbing Wall: Arakawa & Gins Yellow Room
p. 125: Reversible Destiny Sign: Arakawa & Gins
p. 126: Apartment Buildings: Arakawa & Gins buildings
p. 127: Apartment Interior: Arakawa & Gins apartment
p. 128: Yoro Park: Arakawa & Gins YORO Park
pp. 130–31: Man with Surfboard: AdobeStock_290727007

CHAPTER 05: WILL WE BECOME SOFTWARE?

p. 132: Grid over Face: AdobeStock_168579073

p. 135: Goldfish Close-Up: AdobeStock_256426174

pp. 136–37: Model Head: AdobeStock_264597911

pp. 138–39: Photos into Distance: AdobeStock_267651283

pp. 142–43: Hooded Figure, Data, and Face: Free photo 136625222 © Publicdomainphotos–Dreamstime.com, https://www.dreamstime.com/face-blue-beauty-nose-public-domain-image-free-136625222

p. 144: Hand with Brain: AdobeStock_269601824

p. 151: Man on Dock in Data Bubble: AdobeStock_200228257

CHAPTER 06: WHAT WILL IT MEAN TO BE "HUMAN"?

p. 152: Two Cells: AdobeStock_203494644

p. 155: Human and Robot Hands: AdobeStock_201454431

pp. 156–57: Smiling Woman: Free photo 84930844 © publicdomainstockphotos–Dreamstime.com

p. 158: *Born to Be Good* Book Cover: 1001004006849013

p. 159: Two Faces: Free photo 84966424 © publicdomainstockphotos–Dreamstime.com

p. 161: People Picknicking: dreamstime_xxl_84934162

p. 164: Human Hands: human-hands-illustrations-3354675

p. 168: Robot Squatting: AdobeStock_263905049

p. 169: Robot Illustration: AdobeStock_205505599

pp. 170, 173: Woman's Face with White Powder: AdobeStock_25875284

p. 174: Cover of *H+* Magazine: cover-fall-2008 H+ magazine

pp. 178, 181: Particle Face: AdobeStock_256758856

p. 182: Astronaut: AdobeStock_136021997

p. 184: Two Cells: AdobeStock_203494644

CHAPTER 07: ALICE IN FUTURELAND

p. 186: Walking Feet: AdobeStock_39446922

p. 188: Hand Holding Mirror: Free photo under Creative Commons license on Pixel

pp. 192–93: Alice in Futureland Flow Chart: Sputnik Futures

Index

C

D

E

F

G

Q

R

S

T

U

V

W

Y

Z